Shale Gas

A Practitioner's Guide to Shale Gas and Other Unconventional Resources, Second Edition

Consulting Editor **Vivek Bakshi**

Consulting editor
Vivek Bakshi

Managing director
Sian O'Neill

Shale Gas: A Practitioner's Guide to Shale Gas and Other Unconventional Resources, Second Edition
is published by
Globe Law and Business Limited
3 Mylor Close
Horsell
Woking
Surrey GU21 4DD
Tel: +44 20 3745 4770
www.globelawandbusiness.com

Printed and bound by Gomer Press

Shale Gas: A Practitioner's Guide to Shale Gas and Other Unconventional Resources, Second Edition

ISBN 9781787420632
EPUB ISBN 9781787420649
Adobe PDF ISBN 9781787420656
Mobi ISBN 9781787420663

Table of contents

Introduction

Vivek Bakshi
Dentons Canada LLP

In the time since publication of the first edition of this book, the energy industry has continued to operate in a low-price environment with increasing focus on the environmental impact of fossil fuels on climate change. What is clear however is that there remains a role for natural gas under these conditions, notwithstanding the transformational change needed in the energy sector to address the key aims of the Paris Agreement.

This book focuses on shale gas, as it continues to be the fastest growing of the unconventional resources in recent years. Coal bed methane and oil shale are also the focus of some attention, due to the common (but not identical) production, commercial and legal issues involved with these unconventional resources.

What makes shale gas 'unconventional'? The key determining factor is the low permeability of the sedimentary formations from which the natural gas has failed to migrate, and accordingly the different development and production techniques that must be employed as part of exploitation. Other factors that give rise to the description of 'unconventional' include the very different production profile of a shale gas formation and the ultimate rate of recovery from that formation (on both an aggregate and per-well basis). Despite the unconventional nature of the resource and the attendant challenges this presents for producers, the resource potential for shale gas makes these challenges worth overcoming. In its World Energy Outlook 2016 the International Energy Agency projects that in the years leading to 2040, "Natural Gas fares best among the fossil fuels, with consumption rising by 50%".

The sheer scale of potential unconventional natural gas reserves, while significant, is only one part of the picture. Even without unconventional natural gas, most commentators would agree that the size of the world's conventional natural gas reserves is healthy. Perhaps the most compelling case for unconventional natural gas remains how 'equitably' it is spread around the world. The International Energy Agency and the US Energy Information Administration agree that a number of countries and regions which are presently net natural gas importers – or will become increasingly dependent on imports – are precisely those that can now call on reserves of unconventional natural gas. Although this may not be the place to explore the geopolitical implications of one nation gaining energy independence from another supplier nation, energy security remains a key consideration for governments around the world and has fuelled much of the interest in the unconventional natural gas sector in recent years. Add to that the reality that most importing nations have imported natural gas at oil-indexed prices (whether by pipeline or as liquefied natural gas (LNG)), and there is a further financial driver at play if an unconventional resource can be developed economically. Accordingly, in these countries, expect the pace of development of the unconventional natural gas sector to outstrip that in other parts of the world.

In the case of the United States, the most mature unconventional natural gas market, it was declining natural gas production and inelastic demand that raised natural gas hub prices and provided clear price signals for the development and production of the country's shale gas reserves. Those price signals coincided with a fall in development and production costs, due to advancement in technology. It is well known that these same price signals also prompted a significant expansion of LNG import capacity in the United States, but the winds changed following the global financial crisis. The outcome has been that shale gas production, which took off after 2005, has displaced LNG imports and precipitated a remarkable decline in natural gas prices across North America (not only the United States). The impact of the continuing low North American prices for natural gas and the movement to export that natural gas as LNG has had a profound impact on the global LNG market, which is discussed in some detail in this book.

In Canada, where the development of the sector is also progressing rapidly, the key drivers were similar to those in the United States, with one key distinction. Canada has always been a natural gas exporter, with the development of the unconventional natural gas sector designed to help maintain historic export levels to the United States in the light of falling production from the mature conventional fields in the Western Canadian Sedimentary Basin. However, just as growing US shale gas production contracted the market for North American LNG imports, it also contracted the potential market for Canadian unconventional gas; in fact, the traditional demand centres for Canadian natural gas in eastern Canada now import significant volumes of natural gas from the United States (which is often described as Marcellus shale gas), rather than shipping Canadian natural gas cross country from the Western Canadian Sedimentary Basin. To date, holders of significant shale gas acreage in Canada have been unable to move with the speed shown in the United States with respect to the development of LNG liquefaction plants. The reasons for this are also debated in this book. It was anticipated that by 2020 Canada would have at least three LNG liquefaction plants which were to be dependent in large part on shale gas as their feed gas; these predictions are now being reconsidered, with no projects having yet taken final investment decisions.

Poland has shown the challenges presented by shale gas development, despite the considerable potential of the resource. This was arguably the jurisdiction outside North America that most aggressively pursued the development of its shale gas resources, due in large part to reliance on oil-indexed natural gas imports. Development in the sector appears to have stalled in the face of challenges proving resource and a legal and fiscal regime that some commentators suggested was problematic for exploration and production companies.

The use of natural gas as a means of reducing carbon dioxide emissions is a theme that would appear to be consistent with the gradual shift towards a decarbonised global economy. Clearly, as a replacement for other fossil fuels, the potential abundance of unconventional natural gas is precipitating a considerable debate on the climate change virtues of the resource. In countries where coal remains a part of the energy supply mix such as China and in parts of the United States and Canada, an abundant and low-cost shale gas resource will have a role to play. While this is not to ignore the greater carbon intensity involved in shale gas development and production, there is certainly a climate change case to be made by proponents of shale gas development. Where previously there was concern that the supply of shale gas may also cause a contraction in the development of renewable power projects, the better view may be that the installation of significant renewable power generation capacity can and should be balanced by the flexibility and efficiency that natural gas-fired power generation brings.

Natural gas can also play a critical role in reducing emissions from within the transportation sector on a large scale, whether to generate power for electric vehicles or directly in compressed or LNG engines.

The foregoing, and of course the contents of this book, assume that the global shale gas sector in particular will continue in its development and occupy an increasing share of the global energy supply mix, as anticipated by both the International Energy Agency and the US Energy Information Administration. It would, however, be remiss to focus on growth without discussing the environmental responsibility and sensitivity that producers and regulators alike will need to employ to facilitate the anticipated growth in the sector. The industry has already seen that hydraulic fracturing has not received universal acceptance, even in the United States where shale gas production represents a considerable portion of total natural gas production. Bans and moratoria on hydraulic fracturing, while rare, are certainly not unheard of, with France a notable example of a nation with quite significant shale gas reserves which to date has prevented the development of the resource due to environmental concerns regarding hydraulic fracturing. Similar challenges exist in the United Kingdom, with the industry only recently moving back to actively kick-start exploration.

It may be convenient to dismiss environmental concerns as NIMBY-ism – 'not in my back yard' – particularly where experienced developers and first-class drilling contractors are involved in a frack. However, this would appear to be an unfortunate and dangerous strategy. In its 2012 special report, 'Golden Rules for a Golden Age of Gas', the International Energy Agency sums it up well, opining that there is "a critical link between the way that governments and industry respond to these social and environmental challenges and the prospects for unconventional gas production". The environmental challenges referred to are discussed in detail in this book and include the use of freshwater in hydraulic fracturing, the disposal of flowback and produced water, migration of natural gas and fracturing fluids from the fracked formations into groundwater aquifers, venting/flaring of natural gas, and unusual seismic activity.

The industry has moved to address these concerns and groups such as the International Energy Agency and the Canadian Association for Petroleum Producers have prepared best practice guidelines for responsible development. This is clearly to be commended, and notwithstanding compliance by producers with governmental regulation and legislation, there will be a need for public perception to be aligned by the industry with the realities of the production of unconventional natural gas resources. This may be through better education on the practice of hydraulic fracturing and how risks are mitigated by first-class operators. Investment in technology will further mitigate the risks posed by this activity. Although it will not require an unprecedented effort on the part of the industry to align public perception favourably, it will require vigilant and consistent application of these or other best practices as reserves are developed in areas and regions where upstream petroleum development has never been seen before.

Whether you read only a few chapters or the whole book in its entirety, I trust that you will find the contents and the efforts of the authors informative and thought provoking.

The impact of the US shale boom on worldwide natural gas price trends

Nicholas Fulford
Ryan Pereira
Gaffney, Cline & Associates

1. Introduction

In the last five years, the global gas sector – and particularly the more traditional structures surrounding world trade in liquefied natural gas (LNG) – has undergone a series of 'shocks'. North American gas is priced at levels many thought they would never see, especially compared to the relatively high prices of the late 2000s. Exports of US LNG are set to expand rapidly in the coming years, largely a function solely of the significant upsurge in tight gas from shale and similar geological horizons that were not considered viable until recent years. With traditional European and Asian markets either stagnating or shrinking, a new generation of LNG buyers is entering the market looking for low-cost natural gas, largely for gas-to-power projects.

So, how exactly has North American shale gas impacted on the price that is paid for natural gas globally, and what might the future hold?

As we examine the complex interaction of North American gas markets with the rest of the world, it is helpful to describe the effects under four main themes. Each theme contributes to the overall, once-in-a-generation impact, which is beginning to have a transformational effect on global gas markets. The four themes are as follows:

- the economic fundamentals of North American shale gas and why they have delivered a step change in the breakeven price at the wellhead which supports profitable investment;
- the new pricing paradigm which resulted from the early 'tolling contracts' which formed the basis for some 10 million metric tonnes per annum (MMTPA) of contracted supplies to Asia;
- the impact that US LNG exports are having on the global supply/demand balance in the coming years; and
- the effect this could have on accelerating the pace of energy market reform in countries such as Japan, China, India and several emerging LNG-importing countries.

2. How did we get to where we are today – a world where gas is fundamentally priced off oil?

Before examining the current transformation and what may emerge in the future, it is useful to consider how we got to where we are today. LNG has experienced significant growth in the past 10 years, despite the birth of the international LNG industry just over 50 years ago. At its birth, there was no free market for gas; there was a lack of trading options and a lack of liquidity. Oil on the other hand was more established and more freely traded based on supply and demand economics. In addition, the stimulus for liquefying natural gas and transporting it to far away markets that could not be feasibly or economically reached by pipeline was originally intended to substitute other petroleum-derived fuels such as gasoline, petrol, fuel oil, naptha and diesel. The growth and global expansion of gas markets over the past 20 years or so has changed this dramatically. The next few years will see an even greater acceleration of change.

The LNG industry functioned and grew for around 45 years with arguably no impact of US conventional or unconventional gas on internationally traded LNG prices. With the relatively sudden emergence of US shale gas production, the 'golden age of gas supply' may have arrived around five years ago; however, the 'golden age of gas pricing' will most likely emerge in the next five to 10 years.

3. Catalyst for global unbundling of gas/power markets

The interconnection of the two deep liquid gas markets of North America and Western Europe is a major outcome, the implications of which are yet to be fully felt. However, when combined with the general thrust towards unbundled, wholesale energy markets for both electric power and gas, the role of North American shale-derived LNG becomes even more significant.

The US and Canada have the longest history in freely traded natural gas markets supported by a range of exchange-traded contracts which produce a wholesale gas price which is a function entirely of supply and demand. In periods of tight supply, prices rise and encourage additional drilling and exploration. In times of surfeit, prices fall and oil/gas activity subsides to bring gas production into line with demand. These are market fundamentals which have been well proven in other markets, but globally, natural gas has, for various reasons, not followed suit.

The International Gas Union typically considers natural gas to be priced based on eight formulations:

1	Oil indexed	Govern the vast majority (but not all) of long-term gas sale and purchase agreements (SPAs)
2	Gas-on-gas pricing	Applies for most of North American gas, and a large majority of gas traded in Northwestern Europe
3	Bilateral monopoly pricing	Typically involving a national oil company and a state-run utility

continued on the next page

4	Netback arrangement	Where the wellhead price is a function of sales price less the cost of delivery
5	Regulated cost of service	Whereby a regulatory body is allowed to determine the appropriate price payable, often involving a fixed or variable element associated with the costs of running a gas or power utility
6	Regulated social or political	Whereby gas price to various sections of the market differs according to a politically derived formula. For example, selling gas for a lower price in the domestic market, compared to the price for industrial customers
7	Regulated below cost	Still widespread on the African continent, involves a gas price set below the cost of production, using subsidies or other mechanisms to incentivise producers. Sometimes this can be linked with large fines for flaring gas, as another mechanism to encourage production
8	No price	Situations where the gas has no price and is supplied to markets through some kind of obligatory arrangement, completely dislocated from normal commercial practice

Of these pricing mechanisms, the two key ones for LNG are the first two – oil indexation and gas-on-gas competition – although a bilaterally set monopoly price is also present in some regions. The chart on the following page shows a breakdown of gas imports by region, based on the categorisations above.[1]

As we consider the impact of shale-derived LNG, it is useful to consider the history of gas pricing in Europe over the last decade.

4. Lessons from the history of UK and Continental gas market liberalisation

Following the restructuring and unbundling of the UK gas and power industries in the 1990s, a wholesale gas market very similar to that in the United States and Canada started to develop in the United Kingdom. The emergence of a wholesale gas price caused significant financial problems to the incumbent gas utility at the time, British Gas plc. The root cause of the problem was a large, inflexible portfolio of gas SPAs, indexed primarily to oil and other published indices, which generated a price in excess of the prevailing wholesale price. During the mid-1990s a number of retail

1 Source: Gaffney, Cline & Associates analysis.

Figure 1: Portion of gas imports by price category

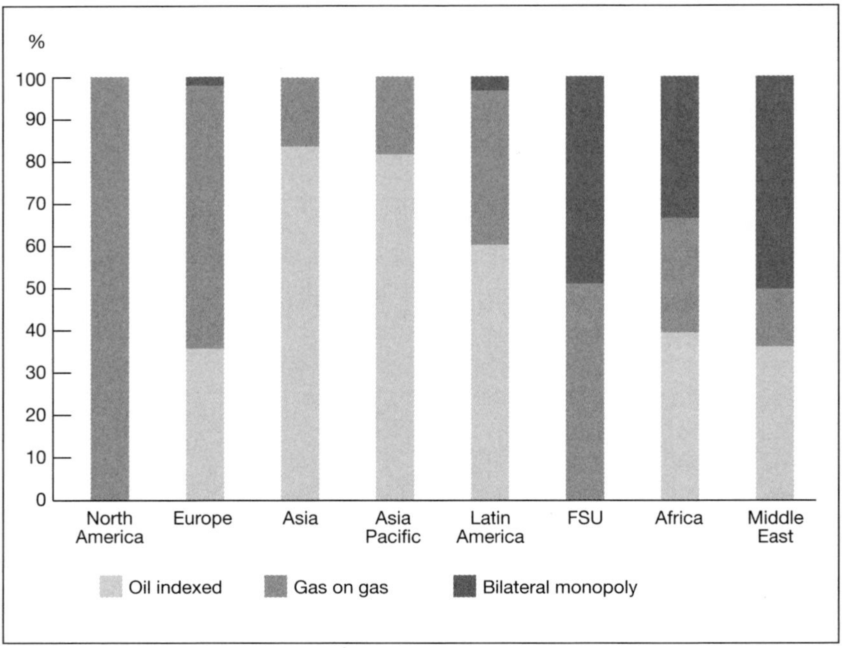

gas suppliers were established to sell gas into the commercial and industrial market, which had been opened to competition as part of the liberalisation of the market. As a result of the ability of these insurgent gas marketers to undercut the retail price offered by British Gas, the incumbent's market share dropped dramatically.

Since most British Gas contracts were take-or-pay, the company found itself contractually obliged to pay for gas it was unable to nominate. Double-edged financial obligations emerged based on the take-or-pay cash liabilities and the fact that the company's weighted average cost of gas was higher than the price at which it could profitably sell to its customers. Official estimates at the time placed a £5 billion post-tax liability on these legacy gas contracts, and a major programme of renegotiations was entered into in order to restructure the company's gas portfolio, at a cost of several billion pounds.

A very similar process unfolded in Continental Europe in the mid-2000s, mainly as a result of two major interconnector gas pipelines between the United Kingdom and Continental Europe. The ability to balance long-term take-or-pay contracts with short- to medium-term traded arrangements based on UK wholesale prices started to have a major influence on how European utilities managed their gas portfolios. Many long-term providers, such as Gazprom and Statoil, found themselves disadvantaged as their take-or-pay contracts were turned down to a minimum while companies profited from the availability of lower-cost, market-priced gas.

This feature, coupled with price reopeners and other dispute resolution mechanisms, led to a series of face-to-face renegotiations of gas contracts, price reopener processes, and litigation and arbitration. This has in turn led, over the last 10 years or so, to a major transition away from oil indexation and the widespread adoption of gas-on-gas pricing. As can be seen from the diagram below, over a relatively short period of time, the basis for gas trading in Europe has reversed from a significant majority of oil indexed pricing, representing around 80% of the market in 2005, to a fundamental change of the situation by 2013, whereby the majority of gas is now traded based on an index such as the National Balancing Point (NBP) or the Title Transfer Facility (TTF).

Figure 2: Migration towards Gas-on-Gas pricing in Continental Europe

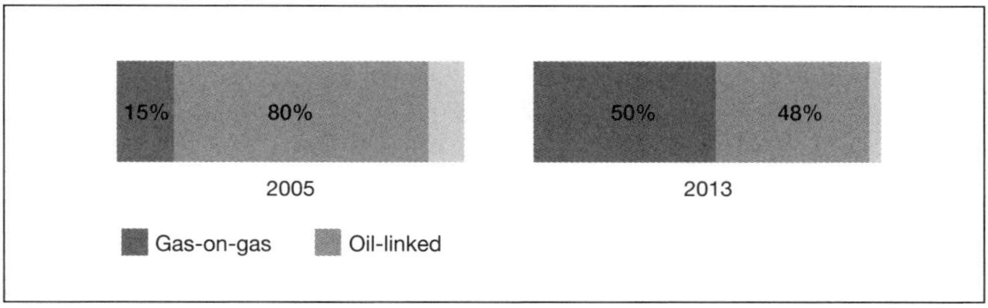

With this assessment of how other markets have responded to wholesale gas competition, we can now return to examining the four main themes that are contributing to North American shale gas having a transformational impact on global gas markets.

5. Reasons for the fundamental step change in low-cost gas production

The first driving factor behind the North American shale gas sector is the development of technology. Technological advances are enabling gas producers to access previously uneconomic source rock, or sedimentary deposits, which while being rich in gas (and oil) lack the permeability of conventional reservoirs that underpinned historical conventional oil and gas thinking. The technologies and related techniques are complex, but the generational change that has seen such success in North America is the ability to use hydraulic fracturing (or fracking) to create a degree of artificial permeability in the source rock to enable the oil and gas to flow.

The remarkable feature of this step change in technology is that not only has it reduced the cost of this so-called unconventional gas, compared to conventional techniques, but over the last few years the cost of producing shale gas has also become less than for many conventional reservoirs. The old paradigm that in the hierarchy of gas sources conventional gas is the lowest cost at the bottom of the pyramid has now been turned on its head. For the foreseeable future, the majority of North American gas development will arise from unconventional/shale gas development.

Based on the current outlook, these low-cost production techniques put many trillions of cubic feet of gas within, roughly speaking, a \$4–\$5/one million British

Thermal Units (MMBTU) breakeven[2] cost zone. The amount of gas in the United States and Canada that can be produced within this cost range is such that it is now more useful to think of the United States and Canada in terms of a 'gas storage tank', where gas can be extracted in whatever quantities are required, over many decades, with a suitable price incentive.

Shale gas or tight gas fall into a category of reservoir that most exploration experts would classify as 'unconventional gas' – that is to say, gas that is produced using different techniques to those that have been in use for the last several decades. Shale gas and tight gas are just two of a number of unconventional sources of natural gas in North America, other examples being coal bed methane and methane hydrates.

Conventional gas production typically involves drilling into sub-surface gas-bearing strata which is relatively porous and permeable, and has, through various geological interactions taking place over millions of years, collected a quantity of natural gas that is trapped. The source of gas for these conventional gas reservoirs is typically the slow degradation of sedimentary deposits, through the action of bacteria or heat that has gradually become oily in nature, before turning to gas. Over a period of millions of years this conventional gas has slowly migrated up to these geological traps (respectively biogenic and thermogenic gas).

Many books and publications are available which provide a more detailed description of that process, but for now we shall put the science aside and examine the main difference where shale gas is concerned. The fundamental change in philosophy for shale, or similar deposits in question, is to access the source rock or sedimentary deposits directly. While this has always been possible, it was not until the 1990s that George P Mitchell famously carried out his pioneering work, largely in the Barnett shale deposit in Texas.

The result is that unconventional gas has become easier and therefore cheaper to extract. To understand the impact that production economics has had on North American gas pricing it is useful to consider the main features which affect the breakeven price of shale gas at the wellhead.

5.1 Type curve

A fundamental feature of any gas well is the anticipated gas flow rate, and the way in which that flow decays as time passes. This can be thought of as a direct proxy for the cashflow anticipated by the producer once the well is drilled, completed and tied back. The characteristics of a shale gas well differ from those of a conventional gas well in a number of important respects.

First, a shale gas well will typically produce at a very high rate in the first weeks or months of its economic life, but in the first few years of operation might decline by perhaps 60%–70% of the previous year's production. At some point, the well will stabilise at a relatively low rate of production compared to the initial flow, and the

2 The breakeven cost is defined as the wellhead netback gas price (referenced back to the Henry Hub through the addition or subtraction of basis differential relevant to the producing basin in question) that will generate a 10% post-tax return for the developer.

annual decline becomes much less pronounced, stabilising at perhaps a 5% reduction on flow year on year. The 'tail' of production can be very long indeed (eg, 10–20 years), but ultimately a point will be reached where the lease operating expense, together with any allocated general and administrative expense, will exceed the cash generated by the gas flow and the well will be plugged and abandoned.

The anticipated initial flow, and the way in which a well is expected to decline, is usually referred to as the type curve, and is a key parameter governing shale gas economics. Although in theory, the type curve for a particular shale reservoir might be expected to be relatively standard, in practice, for wells only a few yards apart the type curve can vary considerably. As a result, data from dozens or even hundreds of wells can be required to determine the true economics of a particular shale play, using statistical patterns from many wells.

One of the areas in which significant progress has been made, but where technology and approach continue to evolve, is in understanding the reasons for variability in the type curve, such that drilling and completion techniques can be honed so as to reduce the number of poorly performing wells and improve average production and decline. Fundamentally, this is what was evolving during the 1990s and early 2000s, bringing shale gas economics within reach of, and then surpassing those of, many conventional wells.

5.2 Well cost

Balanced against the productivity of the well, as defined by the type curve, the cost of drilling and completion is the other key parameter to be considered. For the purposes of a simple analysis, it is useful to divide the cost into two sections:

- Drilling: the cost involved in drilling and casing the well bore; and
- Completion: the cost involved in perforation and fracking.

Drilling costs are largely a result of the amount of time a drilling rig is required to be on station for the initial spudding of the well, drilling of the vertical section, drilling of the horizontal section and the process of casing the well bore to ensure well integrity. In this fashion, other than the long horizontal extensions involved, drilling costs are not markedly different from those of a conventional gas well. However, one area of the drilling process which has received considerable attention is the use of steerable drill bits, which can place the horizontal well bore in the optimum position in the reservoir formation, typically at a location where organic content is high, but the brittleness of the rock enables an effective frack, or cluster of fracks, to be put in place. The accuracy of horizontal well bore placing, sometimes within significantly less than one metre, can have a material effect on well performance.

During the 2012–2015 period, drilling logistics improved considerably through the use of drill pads (which can typically allow for 10 or more wells to be drilled and produced from the same well site), stepping rigs capable of moving to a new location without the need for major dismantling, and better field development planning to reduce the time on station. Producers have also experimented with use of different rigs for the vertical and horizontal part of the well, using a low-specification rig for the first stage drilling and more sophisticated equipment for the more critical horizontal section. As a result of financial pressures on gas producers, by using a combination of these techniques, coupled with lower rig rates generally, significant reductions in drilling costs have been achieved in most US and Canadian shale gas basins since 2010.

Completion costs, on the other hand, are typically an order of magnitude higher for shale gas wells compared to conventional gas wells and in some cases have increased in recent years, but with a more than compensating improvement in the type curve. Completion techniques have evolved since the early days of shale gas production, such that longer lateral sections of wells are utilised and a higher number of perforations and fracks are used to stimulate gas flow. Techniques such as micro-seismic analysis or use of radioactive isotopes in frack sand have been used to better understand the pattern of fracking which the hydraulic fracturing of the rock is producing, and the subsequent correlation with well productivity.

5.3 Oil/gas ratio

However, it should not be forgotten that the fundamental drivers for success are the geological characteristics of the reservoir. One of the key economic factors that govern the profitability of the gas produced at a given market price is the amount of oil, condensate and natural gas liquids (ethane, propane and butane) that are produced from the raw hydrocarbon mixture.

A significant feature of the US and Canadian gas market pricing has not been entirely due to dry gas production, but instead is a feature of the significant volumes of gas that are produced when tight oil is produced.

It was the gas sector that perfected the fracking techniques that have revolutionised hydrocarbon production. However, by the 2012–2014 timeframe it was becoming well understood that the application of fracking to what were previously considered largely exhausted oil basins, such as the Permian or the Bakken, could give oil producers access to substantial, typically underlying formations from which large amounts of oil and condensate could be produced, along with natural gas liquids and natural gas. In fact, the oil economics of some of the emerging US tight oil basins are such that a considerable cost could be borne by the producers to fund disposal of the gas as a waste product. For example, at oil prices in the \$50–\$60 range, a Bakken producer could sustain a cost perhaps an order of magnitude higher than the wholesale gas price and still achieve a profitable rate of return from oil revenues.

Tight oil production in the United States and Canada has had the effect of producing significant volumes of gas which have been sold almost as a bi-product, through the wholesale market, with zero or even negative cost. Although the relationship between oil prices and gas prices is a complex one, a study of market fundamentals tells us that if higher oil prices lead to higher oil production, then they will also lead to more associated gas being placed in the gas market with a potentially countercyclical impact. This therefore has the potential to create a negative correlation between oil and gas prices, which will increasingly start to have effects outside North America.

Bearing in mind all the preceding considerations, ultimately, the gas producer is seeking to create the optimum balance between the initial drilling and completion costs of the well, and the present value of the cashflows that result from the gas produced. Managing this balance is the key feature which differentiates successful shale gas developers from those that have failed. However, the considerable improvement in shale gas economics, driven partly by the potential for higher returns, and also a function of the financial stress caused by low prices, suggests that there may be further improvements available in future. We would expect to see these savings emerge as the industry's understanding of reservoir performance improves and new technologies are tried.

6. New pricing paradigm

The second of the three primary pricing impacts arises from the way in which US gas in particular has been made available to the LNG market. Oil and natural gas markets pre-2013 were relatively well correlated in terms of price. In a segment previously dominated by an LNG price based on a percentage of the oil price, the early US LNG export contracts contained a radically different price determining mechanism. The innovative US mechanism was designed to give the developer of the liquefaction plant a steady and assured rate of return, usually under what is termed a 'tolling contract', with all the gas price risk assigned to the buyer. In such a pricing formula, there are typically four elements to consider when determining the cost of delivering LNG to the customer:

- One of the most material features of that cost is the price of natural gas that enters the liquefaction plant, ultimately to become LNG. Given the highly developed nature of the US wholesale market, this is almost always a feature of the Henry Hub (HH) index, which is the primary price marker for gas in the United States. Therefore, this is the first of the building blocks used to determine the delivered price.
- The pricing will also include a charge to address both the fuel used in the liquefaction train, which leads to a loss of gas between delivery into the plant and onto the ship, and any possible basis differential between HH and the location of the plant. In most of the early tolling agreements, these two features were determined as a 15% surcharge on the HH price adopted. Thus, the HH element of price changes according to the prevailing price in the US market, and is aggregated to 115% of HH, to encompass the two factors above.
- The third element in the pricing, of a similar order of magnitude to the HH index arrangement, is the so-called 'toll' which equates to a lease on the seller's liquefaction facilities. This toll can be seen as a function of the return required by the seller to compensate for the capital investment in the liquefaction trains, storage tanks and marine facilities which are used to liquefy and deliver the LNG. This element of the price formula is fixed and has typically been in the range of \$2.25–\$3.25/MMBTU.
- Finally, there is a freight charge which the buyer must address, which represents the cost of moving the LNG from the seller's terminal facilities in the United States to those of the buyer. Many of the early LNG long-term

SPAs are based on a free-on-board arrangement which means that the delivery point is the flange at the liquefaction terminal in the United States, where the LNG passes on to the LNG carrier, provided by the buyer. This part of the charge is a function of the shipping arrangements entered into by individual LNG buyers, but is often influenced by LNG carrier charter rates, especially for those companies without ships under long-term time charter.

Thus, by aggregating the cost of LNG delivered to the buyer, it amounts to the following:

115% of HH + liquefaction toll + freight to the buyer's regas terminal

The important consideration is that both the structure of the formula, and the quantum arrived at, are very different from the typical oil indexed formula that an Asian buyer might have entered into prior to US LNG export availability.

The attraction to an Asian buyer was therefore threefold:

- mitigating the exposure to global oil prices, which had led to very high LNG prices during the $100/barrel era;
- diversifying LNG price to a commodity with only a weak correlation to global oil prices; and
- offering a much higher proportion of fixed, or less variable features of, price, thereby reducing expected price volatility compared to oil (although arguably S-curves or cap/floor arrangements in LNG SPAs achieve a similar effect).

Arguably, gas prices in the United States and Continental Europe are now close to reflecting the fundamental value of natural gas, rather than a previous reliance on a proxy relationship with oil or other competing fuel. Therefore, as North American shale gas starts to make its way into global markets, the efficiencies of a more gas-on-gas competitive environment will impact global gas pricing, and a global clearing price, influenced largely by regional gas costs and freight charges, will emerge.

7. Impact of increased North American shale gas supply on global LNG supply/demand balance

The third feature of North American shale-based LNG exports is the impact that the volumes produced will have on the global supply/demand balance. In recent times, the most influential suppliers to LNG buyers have been Qatar, where the various liquefaction trains amount to some 77 MMTPA of capacity, and Australia, where by the end of 2017 we can expect to see capacity of nearly 85 MMTPA. Although US LNG exports only commenced in 2016, with the Sabine Pass LNG project in Louisiana, some 60 MMTPA has been approved and is at some form of construction,[3] although of course there is the potential for some of these volumes to be delayed to enable demand to catch up.

Fundamentally, these volumes, coming on top of new Australian capacity, appear to have pushed global LNG capacity and supply well beyond global demand.

3 Source: US Federal Energy Regulatory Commission, US Department of Energy and Cheniere Energy analysis.

Figure 3: US LNG projects

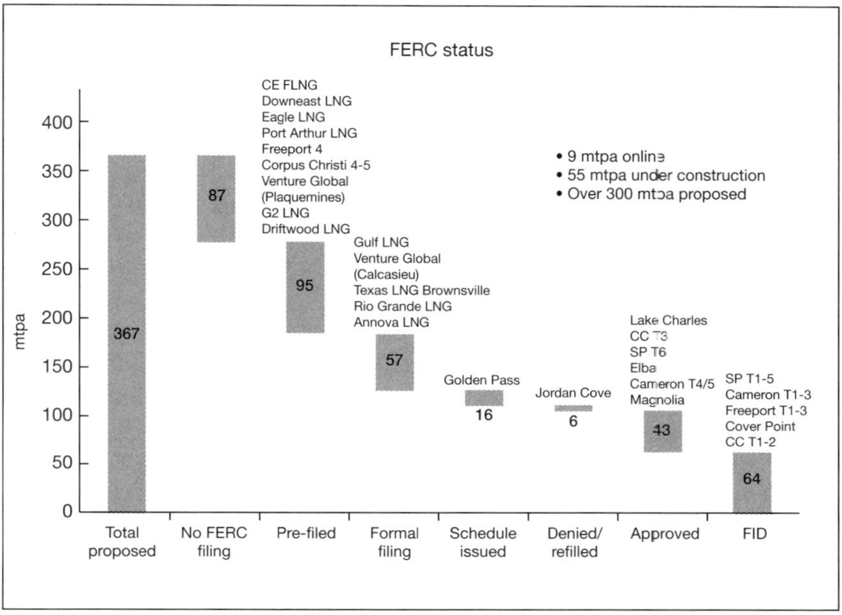

Although approximately three quarters of LNG is subject to rigid long-term SPAs, just over a quarter of traded LNG, in the region of 70 MMTPA, is sold on a more flexible basis, sometimes under a so-called master services agreement, which can apply to a single cargo or group of cargoes. The price of these flexible cargoes is typically negotiated on an arms-length basis between a willing buyer and seller, and therefore is a reflection of market conditions.

The impact of this supply/demand imbalance was felt in 2016, when spot cargoes of LNG were trading in the $5/MMBTU range, although the spot market saw a recovery towards the end of that year, generally anticipated to be temporary. Significantly, the commencement of deliveries of US shale-derived LNG is now creating a tangible link between key global markets, in that much of the gas is in effect globally mobile and will make its way to the highest priced market. This market interconnectivity is already influencing how gas traders approach their business, and in fact new trading products based on this interconnectivity have started trading, including a market in the transatlantic gas price differential, based on the difference between the UK market (based on NBP) and the US market (based on HH).

8. Trinity for change – enabling establishment of liberalised gas markets

In examining the circumstances which led to the establishment of a wholesale gas market in the United States and Canada, the transition that took place in the United Kingdom and the more recent transformation of European gas price mechanisms, three features were present, creating the catalysts for each of these significant market transformations.

Figure 4: Gas Market Liberalisation – 'Trinity for Change'

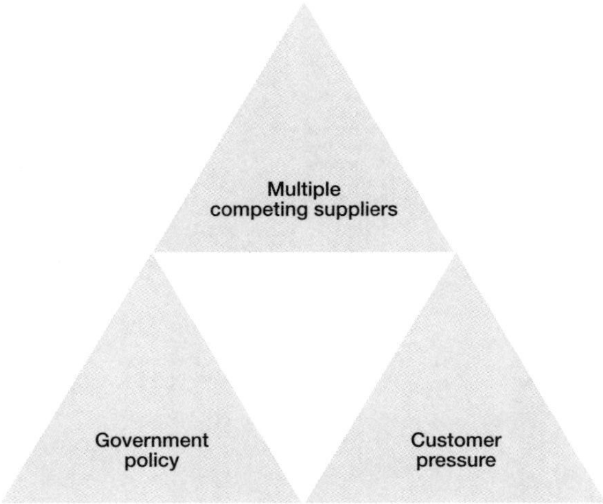

8.1 Competing gas supplies

The key feature underpinning the market transformations has been the emergence of competition in the upstream part of the supply chain, such that, at least with the right regulatory and contractual framework, there has been the possibility for utilities or large gas buyers such as power generators to access multiple supplies of natural gas. This creates an important foundation to support a competitive market, the creation of competition between gas suppliers, and the ability of the lowest-cost provider to capture market.

8.2 Customer pressure

While utilities themselves sometimes create a competitive force, many utilities which have benefitted from cost-of-service regulated prices can be reluctant to bring about an end to that type of low-risk profitability. Change therefore has to be forced upon them by large and sophisticated energy buyers which have a good understanding of upstream gas availability, and which may have had direct discussions themselves with gas providers, planning to use the utility simply as a conduit for delivery, rather than as a gas supplier. In the United Kingdom and Europe, the role that these large energy buyers played in the unbundling/reform process has been considerable.

8.3 Enabling legislation and regulation

The pressure on government from industry leaders is also a significant feature of energy market reform, and the threat of relocating industrial activity to lower-cost jurisdictions is often sufficient to spur government action to put in place enabling legislation and accompanying regulation. The goal of such legislation is typically to facilitate a direct dialogue between gas customers and gas suppliers, through a level playing field arrangement which provides access to LNG reception facilities, high-pressure transmission lines, storage and distribution on a cost-of-service basis.

9. North American shale gas – providing a stimulus for increasing energy market reform in major gas importing jurisdictions

While oil indexed pricing has been historically dominant in the key North Asian markets (primarily Japan, Korea and China), arguably 2015 was a watershed in terms of the emergence of developments to create more transparent market prices that are much more reliant on the supply-demand fundamentals of the commodity being traded – natural gas.

When looking at these historical developments against the impact that shale-derived LNG is having on global markets, it is important to consider some of the key markets and how they are addressing the question of energy market reform and unbundling.

9.1 Japan

Currently accounting for just over one third (37% in 2014) of the global LNG market, events in Japan are of crucial importance to the overall sector. By 2015 some 10 MMTPA of LNG imports had been contracted on a tolling basis through the various US terminals, although at that time, no LNG was actually flowing. Japanese LNG buyers were looking to North American shale gas to deliver portfolio diversification. Japanese LNG buyers were also investing further up the value chain in North American shale gas plays, investments that have in hindsight not always paid off given the subsequent collapse in natural gas prices. However, in the longer term they could serve as a natural gas price hedging policy.

Although the process of Japanese interest in US LNG had started prior to the earthquake in 2011 and the subsequent shut down of the Japanese nuclear stations, the measures taken to improve the resilience of the Japanese gas and power market at that time accelerated the process. In particular, Prime Minister Abe launched a major legislative programme under the heading of Energy Market Reform, a large part of which was to dismantle the monopolies enjoyed by the major gas and power utilities, and introduce a competitive market place similar to that which had emerged in Europe. This reform started with the opening up of the power markets in April 2016, with the gas market to follow in April 2017. The fragmented nature of the gas and power grid in Japan creates various additional hurdles which mean that the gas and power markets may take some time to develop the flexible dynamic nature of the markets in North America and Europe, but nevertheless have significant implications for LNG buyers and sellers.

The emergence of new entrants licensed to sell power and gas to a variety of customers means that some degree of competition will emerge in the downstream markets, which will largely be price driven. A new entrant, able to capitalise on the anticipated LNG oversupply over the next few years, might therefore be expected to be able to offer a lower energy price than one of the established utilities, whose weighted average cost of gas is largely determined by legacy oil contracts. As a result, the pace of change in Japan could well be quicker than most market commentators are perhaps predicting.

In this way, the same situation that occurred in the United Kingdom and Europe may develop, and LNG buyers may put significant pressure on their suppliers to adopt a more market-oriented (gas-on-gas) price. Prior to the oil price realignment of 2014–2015 there were signs that Japanese buyers were putting pressure on their suppliers, either within the price reopener clauses, where appropriate, or simply through other means to renegotiate their oil-linked portfolios. Although this pressure has eased, as the price of oil indexed gas, indexed (HH) gas and spot cargoes continue to adjust to market conditions, it can be expected to resume, once some degree of stability returns to oil and gas commodity pricing, and a global clearing price for gas becomes a better-established concept.

9.2 China

Although currently not yet as influential in LNG markets as the traditional buyers of Japan and Korea, China has the potential to play a pivotal role in LNG pricing over the next decade. The availability of significant quantities of shale-derived LNG from the United States, at a time when this potentially huge gas market is evolving, is likely to be one of the determining features of global gas pricing over the next decade.

The Chinese gas market is characterised by significant support, as a result of the following factors:

- natural gas becoming a major plank of Chinese environmental policy aimed at reducing the damaging pollution in most Chinese conurbations caused by a blend of particulates and sulphur dioxide and nitrogen dioxide emissions from coal, fuel oil and diesel burning;
- a regulatory environment which is seeking to price gas competitively in comparison to competing fuels such as fuel oil and liquefied petroleum gas;
- a significant initial foray into LNG purchasing through the construction of major regasification terminals by the three state-owned players, CNOOC, CNPC and Sinopec; and
- a second phase of major investment funded by power generators and other gas users who are frustrated by the lack of progress in achieving third-party access to existing LNG regas facilities, many using the planning delegations which enable provincial approval only of regas facilities of less than 3 MMTPA.

While the reluctance of the state players to facilitate third-party access is understandable on some levels, it may create some take-or-pay issues for the companies involved which have signed long-term SPAs based on conventional oil price indexation.

The second-tier companies are rapidly building additional LNG regas and pipeline infrastructure to provide low-cost gas-fired generation, which has resulted in significant regas spare capacity similar to the situation in Europe, but with the difference that Chinese gas demand is growing. If these second-tier companies are able to negotiate SPAs based on the emerging paradigm of gas-on-gas competition, and the power market also moves to a competitive arrangement driven by merit order, as in North America and Europe, those same take-or-pay problems of uncompetitive gas could conceivably start to affect Chinese importers. Indeed, there are already signs that Chinese buyers are seeking to renegotiate their LNG SPAs to achieve a more competitive price formulation.

Recent market developments suggest that for price-sensitive Chinese buyers, the formulation of some kind of quasi-global clearing price for gas has already emerged in some recent long-term LNG SPAs. The concept adopted uses a number of regional market indicators, particularly HH, NBP and Japan Korea Marker, together with various constants which are designed to approximate to the freight cost of moving gas from one market to another. This methodology leads to a hybrid formulation designed to protect the buyer from a gas price that becomes dislocated from what other buyers in the region may be paying. We can expect to see this type of sophisticated price formulation become more finely tuned as regional gas markets become more interconnected, mainly as a result of globally mobile shale-derived LNG, and prices become more transparent.

Figure 5: Emerging Hybrid Pricing Trends in LNG SPAs

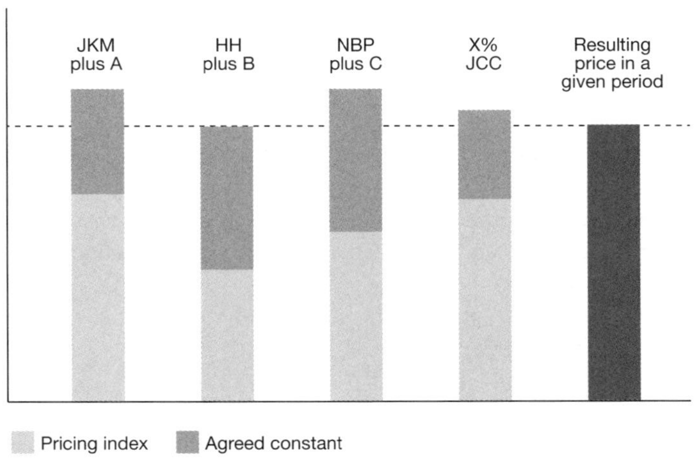

Constants A, B and C are negotiated, and based on typical price differentials which may reflect average transportation costs from a basket of LNG suppliers.

In such a way, the impact of North American shale-derived LNG could become one of the major catalysts for price reform in China. However, the outcome of this process is hard to predict due to the complex relationships that exist between central government, gas and power price regulators, state-owned and controlled entities, and private investment.

9.3 India

The Indian gas market has proven the most price sensitive of all the major emerging LNG buyers, and there is also evidence of a much more active price dialogue which has significantly rebased a number of the LNG contracts that have been entered into, such as LNG deliveries from Qatar.

If shale-derived LNG from the United States starts to create clear linkages between markets, we can expect to see Indian LNG buyers becoming early adopters and looking for price advantage where possible.

9.4 Other Asian markets

As we approach 2020 we can expect to see quite a number of the emerging LNG markets in Asia adopting some form of energy market reform in a bid to establish a price mechanism for gas which sets an appropriate pricing signal for power generation and industrial markets.

Thailand has an unbundling regulatory process which is set to unfold in the next few years, and Vietnam has also set out a longer-term objective of a functioning wholesale market for gas, broadly modelled on the power wholesale market that is already emerging.

Malaysia, the Philippines and Indonesia all have price regulation and market reform processes on the go, and as more market-priced LNG becomes available, we can expect the process to accelerate.

There will be an amplified effect given significant gas import infrastructure considerations in Bangladesh, Sri Lanka, Myanmar and Vietnam.

10. What may emerge in international LNG markets over the next decade?

At present, we have in essence three competing price mechanisms:

- the standard long-term take-or-pay contracts, based on oil;
- there is now more than 10 MT of LNG signed up on some form of HH-linked price; and
- a growing portion of the global LNG volumes that are traded on short- or medium-term contracts.

If Asian, and indeed global, gas prices follow the progression we have seen in North America, the United Kingdom and Continental Europe, what may gradually occur is a merging of these pricing mechanisms into some kind of hub or potentially multiple hubs addressing key markets. This will be supported by the significant new flows of LNG currently coming to the market, predominantly from Australia and emanating from North American shale gas-derived LNG. Hub-based gas-on-gas pricing reflecting global supply/demand may emerge at a quicker pace than many are predicting.

11. Conclusions on the impact of North American shale gas on global LNG pricing

The impact of shale-derived US LNG is already material, and will become more so over the remaining years of the current decade and into the early 2020s. There is no doubt that the transformational economics of shale gas, garnered largely from significantly improved productivity from fracking and related completion techniques, as well as improved drilling logistics and related cost savings, will continue to be honed. Additional technologies are emerging day by day which are

Figure 6: Potential pricing model over the next decade

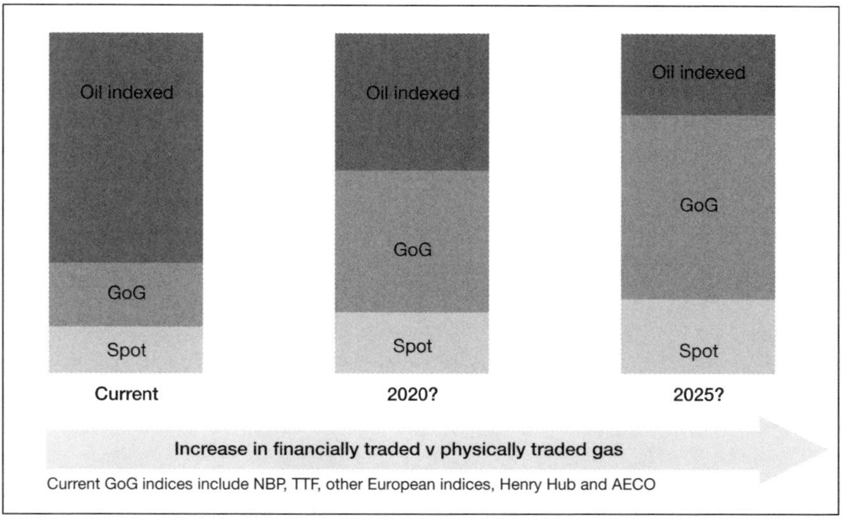

helping producers to bring poorly performing wells up towards the standards achieved by the top-quartile performers. In doing so, more contingent reserves will be added, underpinning the role of North America as the world's 'gas storage tank'.

US LNG producers are likely to remain at the forefront of price formulation, with hybrid pricing, potentially based on a range of global gas indices, taking a firmer hold in LNG SPAs. This will gradually give rise to a global clearing price mechanism for LNG that can become an increasingly dependable and financeable feature of the market.

It seems highly likely that the US liquefaction projects under construction will cause a medium-term excess of LNG supply over demand which will continue to encourage new LNG importers, many of whom may be prepared to move away from the traditions on which LNG SPAs have been based for many decades. As this process unfolds, the establishment of a functioning wholesale market for natural gas, supported by a range of exchange-traded contracts and financial players, is likely to mean that the market mechanisms that we have been used to in North America and Northwestern Europe for many years will take hold in new markets. In this respect the energy market reform programme in Japan, and the steps to create a more market-oriented gas price in China, will be key developments to watch. Oil indexation in LNG pricing will continue to play a role, but could quickly die a death as LNG priced on gas fundamentals emerge.

The story of shale-derived LNG arising from the United States, and the shift in pricing levels and paradigms in global gas markets, has yet to be fully written, but all the indications are that we are set for a seismic event in LNG, with the United States at its epicentre.

Gas sales agreements and natural gas trading

Jessica Davies
Clifford Chance
Steve Delic
BP Exploration and Production

1. Introduction

The global gas industry is experiencing a revolution, with abundant new sources heralding a profound shift in the balance of market power. Increasingly, unconventional sources of gas (those that have traditionally been seen as difficult and expensive to commercialise), including shale gas (gas trapped in shale formations, often found close to coal deposits), have become an important part of the story.

The recent shale gas boom in the United States resulting from major discoveries and advanced technology – in particular, increased use of and enhancements in hydraulic fracturing and horizontal drilling – has unlocked vast tracts of gas-bearing shale and created a global game-changer. Other factors have been critical to the US success story, including access to existing infrastructure, a vibrant services industry and favourable tax concessions. This dramatic shift in available natural gas reserves in the United States has converted it from a liquefied natural gas (LNG) importer to an LNG exporter in a relatively short period of time, and has drastically reduced its energy dependence – domestic shale is predicted to account for one third of US energy production by 2020, and the US Energy Information Administration projects that the United States will be a net natural gas exporter by 2018. The prospects for LNG exports continue to drive new projects: the Sabine Pass LNG export plant, the first of its kind to ship US shale gas overseas, made its first shipment in early 2016, and the Corpus Christi terminal in Texas, among others, has come online.

However, the shale boom has, among other things, precipitated a global gas glut, leading to significantly depressed prices. For instance, in Australia, where a number of mega-projects are being developed, LNG exporters are facing pressure from both a rise in US shale gas exports and the potential of China's shale gas reserves. A further threat is the intense local competition for the resources required to construct and operate these mega-projects. Moreover, there is uncertainty surrounding the amount of recoverable shale gas that can be economically extracted outside of the United States.

Shale gas developments are challenging because:

- they can have higher production costs per volume than conventional gas;
- their decline rates appear to be more accelerated than conventional gas;
- they raise environmental concerns around fracking, leading to increased public scrutiny and environmental regulation; and
- experience with and access to extraction technologies is required.

The level of capital investment required for shale gas exploration, production and distribution would suggest that long-term gas supply agreements (GSAs) are needed to underpin the investment. These arrangements typify the traditional LNG export market through long-term LNG sale and purchase agreements (SPAs), while a mix of varying term agreements and short-term trades characterise the domestic market. However, the shale gas revolution in the United States has been notable for its increased use of short-term GSAs, swaps and hedging strategies, given it is a buyer's market, and they are increasingly looking towards more bespoke structuring to take advantage of the market, including to optimise shipping and pricing arbitrage opportunities.

This chapter focuses on the motivations that drive the use of short-term GSAs over the more traditional long-term GSAs, and provides a high-level overview of the key elements of long-term GSAs and short-term trading. The United States and Australia are used as case studies to demonstrate how various factors underpin trading and gas supply decisions and structures. Traditionally, in a mature domestic gas market such as the United States, which has a developed secondary trading market, short-term GSAs, swaps and hedging are useful strategies to mitigate the volatility of natural gas prices and investment. In contrast, Australia – with a burgeoning conventional and unconventional gas industry, immature natural gas secondary trading markets and unique geographical challenges (eg, its disconnected regional gas markets) – has historically promoted the use of longer-term GSAs. However, substantial developments in the LNG export market in Queensland have recently increased the need for more flexible transactions between parties, and have caused a significant increase in the amount of short-term LNG trading. Since many long-term contracts are linked to oil, the recent drop in oil prices has had pass-through effects on LNG prices in the large markets of Asia and Europe. Combined with lower-than-expected global demand growth, this has driven down both oil-linked contract and spot prices, narrowing the price differentials between the Americas and Asia-Pacific.

2. US approach to gas sales and trading

Domestically, the US natural gas market has historically been affected both by geography and the distances over which natural gas must be transported, which play a large part in determining natural gas prices. Before the peak of the shale boom, natural gas was sold under long-term GSAs due to a relatively stable market in which the incremental cost that such distances caused was predictable. However, following the discovery of shale gas, the market has become increasingly volatile. This is due to a number of factors that make long-term pricing less attractive in favour of short-term instruments that can adapt to ever-changing market conditions – for example:

- the ever-increasing estimates of how much natural gas is available in the US shale plays;
- the pace at which natural gas explorers are accessing that natural gas;
- the many supply locations that are coming online and shifting the physical market's dynamics;
- the paradigm-shifting effect of the sustained low price of natural gas, which

is leading to changes in energy infrastructure requirements in the United States; and

- speculation around investment in the long-term outcome of such a paradigm shift to preferred energy sources and technology, leading to factors such as climate change, environmental regulation and renewable technologies playing a greater role than ever before.

Traditionally, the location of a natural gas purchaser was an important and predictable factor in the pricing of natural gas, which led to a stable pricing market and the use of long-term agreements. The physical market for natural gas is in southern Louisiana (the Henry Hub); if the buyer was located in Chicago, the transport price from Chicago to Louisiana was a predictable cost (the basis differential) that could be added to the price of the contract. Utility companies or other distributors, as relatively unsophisticated players, relied on long-term agreements, as simple pricing structures based on distance enabled predictable costs over extended periods.

In terms of exports, until recently the United States has largely exported natural gas via a pipeline to Canada and Mexico, with a small volume of LNG exported to Japan (from Alaska); as previously noted, this has and will continue to expand dramatically with the conversion and construction of export facilities such as Sabine Pass. The expansion of the Panama Canal has also seen a new trade route opened from July 2016. The shorter trade route between the Atlantic and the Pacific is now accessible by more of the existing global LNG fleet and will help to cut costs, resulting in greater volumes of trade as incremental cost margins improve. This improved trade route, together with a rapid increase in US supply, is gradually making the prospect of a viable LNG export industry a reality.

Technological developments, including hydraulic fracturing, have been a key factor in unlocking the numerous shale plays in the United States and transforming the US energy market. From 2009 to 2015, the Environmental Protection Agency's estimates of the total volume of natural gas contained in US shale plays were continually revised upwards. According to Energy Information Administration estimates, US petroleum and natural gas production first surpassed Russia in 2012, and the United States has been the world's top producer of natural gas since 2011 and the world's top producer of petroleum hydrocarbons since 2013. Energy companies have wasted no time in investing resources in these new technologies. Initially, this new 'shale gale' was dominated by smaller energy companies which focused on lease acquisition by gaining landowners' permission to access the natural resources, including gas, underneath their properties. This was a major undertaking, as the majority of the natural gas was underneath small, privately owned land parcels and access had to be negotiated with individual landowners. In some cases, landowners created groups to leverage more bargaining power with the energy companies, but this tended to slow the process down rather than making it more efficient. Following this initial wave, larger energy companies began to buy lease rights from smaller companies, either capitalising on the drained capital and lack of income of the smaller companies, or simply valuing the total assets on the secondary market. These bigger companies were well positioned to begin construction and drilling operations immediately, as they often had excess capital on hand. During this period, general awareness of shale gas in the United States began to increase, generating discussion about its environmental impact. The Marcellus Shale in New York, Pennsylvania and Ohio – one of the largest US shale plays – is in close proximity to the biggest natural gas-consuming states. This lowers the price of natural gas overall, since the biggest consumers are no longer those situated farthest from the source. However, it also shifts the dynamics of the market, allowing much more robust competition for both suppliers and consumers in the marketplace. This also has the effect of depressing natural gas prices overall.

In addition to the various physical market factors, several longer-term factors are contributing to market volatility. One major effect of the sustained low price of natural gas is to spur investment in infrastructure that utilises the new network locations and technologies of natural gas. Examples of such investments include:

- developing pipelines from these new sources to existing consumers;
- converting facilities designed to import LNG to facilities designed to prepare natural gas for export as LNG (eg, the Sabine Pass facility); and
- retrofitting existing technologies (eg, power plants and engines) to use natural gas for energy production.

The speed of investment in these conversions and technologies affects the demand for natural gas, and in turn the price; therefore the rate of investment is subject to speculation. This speculation occurs not only in the physical market (between producers and consumers), but also in the financial market that trades in natural gas as a commodity. Finally, other factors which affect the demand for natural gas (eg, the shape and extent of eventual environmental regulation of shale gas in the United States; investment in renewable technologies such as solar and wind power; and the market for natural gas exports as other nations begin to produce shale gas) all lead to volatility that favours short-term trading instruments over long-term GSAs. If anything, long-term GSAs represent a financial pitfall for energy producers at present, as the current market price for natural gas is extremely low. Many shale gas producers are facing capital constraints due to these low prices – a condition which an LNG SPA would only exacerbate. In addition, if producers sign long-term agreements at such prices, they significantly reduce the value of their main asset – their newly unlocked natural gas. This affects the producer's profit margin as well as the attractiveness of the company as an acquisition target. As previously mentioned, many small producers entered the market early and were then acquired by bigger players – a strategy which an LNG SPA at today's low natural gas prices would be likely to jeopardise.

3. Australian approach to gas sales and trading

As explained further below, in Australia all LNG is exported. The domestic gas market comprises three distinct regions:

- the large eastern market (comprising Queensland, New South Wales, Victoria, the Australian Capital Territory and Tasmania, interconnected by a network

of transmission pipelines and principally supplied by the Surat-Bowen, Cooper, Gippsland and Otway basins);
- the western market (Western Australia) supplied by the Carnarvon and Perth basins; and
- the northern market (Northern Territory) supplied by the Bonaparte and Amadeus basins.

Interestingly, despite continued depressed pricing worldwide, pricing in parts of Australia remains volatile. Gas producers and pipeline operators on the east coast in particular have been operating in an industry undergoing significant change in the last few years: big energy producers have been rapidly building massive capacity for the export LNG market out of Queensland. Industrial users have received fewer offers when seeking to buy gas, at higher prices, for limited volumes over shorter periods of time and on more restrictive terms and conditions. This has ultimately resulted in deferred investments, shorter-term GSAs or reduced gas usage.

This prompted an inquiry by the Australian Competition and Consumer Commission (ACCC) in 2016. In its "Inquiry into the East Coast Gas Market", the ACCC found the main issue affecting supply was uncertainty regarding the future supply outlook and a lack of transparency and information on reserves, resources and prices. Pipeline operators have responded by investing in new capacity; however, the Australian Energy Market Commission and the ACCC are concerned about monopoly pricing and withholding capacity on certain regional pipelines. Monopoly pricing gives rise to higher delivered gas prices: the ultimate cost borne by consumers. Some of the key factors the ACCC found affected the market included the magnitude of gas flows to LNG projects removing gas from domestic markets and the low oil price, including declining investment, uncertainty about future supply, and regulatory restrictions, including moratoria on fracking.

3.1 Shale potential

With geological and industry conditions resembling those of the United States and Canada, Australia has the potential to be one of the next countries with commercially viable shale gas and shale oil production. Geological information indicates that six key basins in Australia hold an estimated 2,046 trillion cubic feet (TCF) of risked shale gas in place, with 437TCF as the technically recoverable shale gas resource. Of the six assessed basins, the Cooper Basin, Australia's main onshore gas-producing basin, with its existing gas processing facilities and transportation infrastructure, could be the first commercial source of shale hydrocarbons. Santos, Beach Energy and Senex Energy are testing the shale reservoirs in the Cooper Basin, with initial results from vertical production test wells providing encouragement for further delineation.

While Australia's shale gas reserves may be technically recoverable, this does not guarantee commercial viability. In fact, many experts estimate that any significant production of shale gas in Australia is still years away. Small independents have led the way in assembling the geological data and exploring the high-potential shale basins of Australia. International majors are now entering the market by forming joint ventures with smaller independents, bringing capital investment. However, several of the key factors which make shale gas a viable commercial proposition in the United States (eg, access to pre-existing infrastructure, the separation of capacity rights and ownership, and the support of a large and competitive services industry) are not present in Australia. Many of the potential Australian shale plays cross state and territory boundaries, which could present further hurdles for a producer due to the lack of national regulation and varying levels of state regulation. With the remoteness of many of Australia's shale gas and shale oil basins, development will likely proceed at a moderate pace.

4. Contrasting motivations of long-term and short-term interests

The aim of a long-term GSA is to secure a viable market for the gas before commencement of the project due to the complexities and upfront capital investment required. Indeed, proponents will need to demonstrate that the project is commercially viable by demonstrating to potential investors that a secure and reliable market for the natural gas exists. Traditionally, the dominance of long-term GSAs was a consequence of the need for both producer and seller to ensure adequate opportunity to recoup their investment. From a purchaser's perspective, a long-term GSA guarantees the security of long-term supply, but also limits flexibility, locking it into a long-term purchasing commitment based on a given pricing methodology. A long-term GSA will seek to balance the competing motivations of the seller and the purchaser, and achieve a balance between the requirements of the project and the need to provide the purchaser with a commercially acceptable price.

In contrast, spot trades and short-term GSAs are focused on the immediate needs of the purchaser and seller, and provide much greater flexibility. The parties are less concerned with being able to address or mitigate issues that could arise over a period of time, and more concerned with executing the transaction quickly. Short-term GSAs, and particularly spot trades, were initially used to dispose of excess gas or acquire gas to cover shortfalls, and to supplement the relatively inflexible long-term contract market by providing greater efficiency and liquidity and increasingly greater arbitrage opportunities.

5. Long-term gas sales agreements

Long-term GSAs have traditionally been the predominant form of sale agreement for LNG exports (LNG SPAs), and they are typically heavily negotiated between the parties. Unsurprisingly, in the current market they are increasingly on 'buyer friendly' terms.

This section provides a high-level overview of key issues that are generally present in traditional LNG SPAs, in particular:

- conditions precedent;
- start date;
- duration;
- quantities and scheduling;
- take-or-pay and make-up provisions;
- *force majeure*;

- liabilities;
- pricing; and
- termination issues.

In addition, the section examines some of the alternative regimes that are becoming increasingly prevalent in LNG SPAs. Long-term GSAs used for domestic or pipeline gas are on similar terms (ignoring LNG-specific provisions around cargoes, shipping etc).

5.1 Conditions precedent

The conditions precedent in an LNG SPA vary considerably from contract to contract as they depend upon the particular structuring of that project. As LNG SPAs generally pertain to complex project structures, often the ancillary contracts related to the project cannot be completed simultaneously. Consequently, an LNG SPA is often the first agreement to be completed and the execution of subsidiary agreements will be conditions precedent to the commencement of the LNG SPA. Other common conditions precedent include: financial investment decisions by one or both parties; proof of reserves sufficiency; third-party arrangements (either upstream or downstream); finalisation of financing for the project; and relevant consents, including government approvals, permits and licences.

5.2 Start date

The start date of an LNG SPA is the date on which the parties' obligations under the contract commence. This is generally the first delivery date for the seller and the date on which the take-or-pay obligations commence for the purchaser. Often the start date is several years after the effective date of the GSA as a consequence of one or both of the parties developing facilities to enable them to perform under the contract. The start date can be a fixed date, but more typically is determined by a window-narrowing mechanism which enables the parties, by written notice, to shorten progressively the period of time within which a party may nominate the start date.

5.3 Duration

LNG SPAs have generally ranged from 15 to 30 years in duration; the long duration of the agreement reflecting the substantial capital investment required to facilitate the operation of the project. The contract term is generally measured from the first delivery date, and the obligations to deliver and take or pay traditionally end at the expiry of the contract. There may be a degree of flexibility with respect to the contract duration to reflect the reserves capacity uncertainty; often, however, the insertion of a declining production tail in the GSA resolves this issue.

5.4 Quantities and scheduling

The quantity and calorific value of gas that the seller is to provide and that the purchaser will buy should be set out in the LNG SPA, and is typically defined in terms of the overall contract quantity, the annual amounts and, if necessary, more frequent measures of quantity.

The overall quantity of gas to be delivered under the LNG SPA is clearly the first quantity to be determined by the seller and the purchaser. The seller typically calculates the amount of gas that it is ready to provide to the purchaser by reference to the reserves available and previously contracted volumes to other purchasers. Security of supply is a key issue for purchasers, which have traditionally insisted on third-party expert certification with respect to sufficiency of reserves, and have often required seller undertakings not to oversell to third parties. Purchasers of shale gas must be aware of some of the idiosyncrasies of shale gas reserves, including their shorter production life compared with conventional reservoirs, the uncertain nature of recoverable reserve estimates due to the immature shale market, and potential environmental issues interrupting production. These factors may make the task of estimating the overall contract quantity more difficult than determining the economically recoverable reserves of a conventional gas reservoir.

In addition to the overall contract quantity, an LNG SPA will set out the annual contract quantity. The annual contract quantity represents the notional maximum amount of gas which the seller will deliver to the purchaser each year under the contract. It may be determined purely by reference to the overall contract quantity, or it may be derived by multiplying the monthly or daily contract quantities. Often an annual contract quantity provision will provide a degree of flexibility to adjust the stipulated amount through the inclusion of a mechanism to adjust the overall contract quantity in certain circumstances. Upward quantity flexibility enables the purchaser to receive more gas than is provided for in the annual contract quantity in a given year up to a stipulated amount, whereas downward quantity flexibility allows the purchaser to reduce the annual contract quantity in a given year, again to a stipulated amount. The degree of flexibility that the seller will be prepared to concede to the purchaser is highly dependent on existing market conditions and varies from contract to contract. Clearly, a seller would prefer to limit a purchaser's downward flexibility; where such flexibility is allowed, a seller will attempt to ensure that the purchaser is required to 'make good' the quantity which was not delivered in another contract year, particularly in jurisdictions where secondary trading on the spot market is limited. If the annual contract quantity provisions include a make-good obligation, the parties should ensure that the time period to enforce the obligation is certain (eg, will the obligation terminate at the end of the term of the agreement or will it continue after the term?), and that the conditions surrounding the exercise of the obligation are clear (eg, will the make-good quantities be in addition to the purchaser's annual contract quantity amount, and does the buyer have to ensure that the make-good quantities are available for the buyer, or are only reasonable endeavours required?).

5.5 Take-or-pay and make-up provisions

Take-or-pay provisions are commonly included in LNG SPAs to provide the seller with further reassurance that the purchaser will accept delivery of a minimum quantity of gas or otherwise pay for the minimum quantity of gas. In general, the annual contract quantity acts as the cornerstone for the take-or-pay quantity, with many contracts setting the minimum quantity at close to 100% of the annual contract quantity. The insertion of such a provision is motivated by the seller's desire to recoup its investment and to increase the gas project's bankability as a consequence of the guaranteed cash flow. In effect, the seller is obliged to supply the purchaser with the annual contract quantity of gas, but the purchaser is not obliged to take it. Instead, the purchaser is obliged to take and pay or not take and still pay. To avoid a potentially onerous take-or-pay obligation, a purchaser must determine the circumstances in which the take-or-pay obligation will not apply. Circumstances to consider including are: *force majeure* events (discussed below); a seller's failure to deliver in the explicit delivery period; and the delivery of gas which does not meet the agreed specifications in terms of volume and/or calorific value.

The negotiation of take-or-pay provisions often involves considerable time as, regardless of market conditions, sellers are reluctant to guarantee the make-up quantities, since this would require them to reserve capacity for the purchaser in each contract year where they have a make-up entitlement. The make-up provision acts as a counterbalance to the potentially oppressive nature of the take-or-pay provision. It enables a purchaser to 'bank' the quantities that it does not accept under the take-or-pay provision to use as credit in a future contract year. Both the seller and the purchaser should again determine the circumstances under which a make-up obligation will arise, and the requirements imposed by that make-up gas obligation.

5.6 *Force majeure* risk

LNG SPAs are commonly governed by the laws of England and Wales, or by New York law. As neither legal system provides for *force majeure* relief, *force majeure* provisions are often included in LNG SPAs in order to provide that, where an event or circumstance beyond the control of either party causes a party to fail to meet its obligations, that party is relieved from liability. A *force majeure* provision, while relieving a party from liability, should not suspend the party's obligation under the LNG SPA. Instead, the provision should expressly provide that such a party should continue to attempt to fulfil its obligations under the GSA using reasonable endeavours.

As *force majeure* is not a term of art, it is a key area of negotiation between the parties, and as such it is important to define the term with accuracy and in detail. As a consequence of their frequent use in LNG SPAs, *force majeure* provisions include the following common features:

- a detailed definition of a *force majeure* event;
- examples of events which will constitute a *force majeure* event;
- obligations surrounding a *force majeure* event; and
- a specific list of events which will not constitute a *force majeure* event.

5.7 Liabilities and indemnities

An LNG SPA may not explicitly outline the allocation of liabilities between the parties with respect to loss or damage to each party's personal property or interests arising from activities contemplated under the LNG SPA. If this is the case, liability will be allocated according to the laws governing the agreement. For example, under English law, a party that causes loss or damage to the other party's personal property or interest will be liable if liability is established. However, LNG SPAs have developed fairly standard liability and indemnity regimes which specifically set out the consequences of liability. The two following models are typically utilised in LNG SPAs:

- the 'mutual hold harmless' model, under which each party assumes responsibility for loss or damage to their personal property or interests, regardless of whether the other party was responsible by act or omission for such loss or damage; or
- the 'fault-based' model, which is a contractual reiteration of the English law position, under which a party which causes loss or damage to the other party's personal property or interests is required to indemnify the other party for such loss or damage.

Another key area of negotiation in an LNG SPA is that of the seller's liability with respect to a failure to make available or deliver gas, or the delivery of off-specification gas. Generally, an LNG SPA will provide for damages or liquidated damages in the event of seller shortfall. However, the extent of a seller's liability with respect to a purchaser's direct and consequential damage varies considerably between LNG SPAs. A seller will not generally be held liable for any consequential damage which flows from a seller's shortfall. The only exception which is generally explicitly provided for in LNG SPAs is where a seller's wilful misconduct causes the shortfall. In such circumstances a seller is generally held liable for a purchaser's direct and consequential damage. The law of contract is often inadequate to protect a purchaser sufficiently. To hedge against the possibility that the law of contract will not provide adequate protection, a purchaser could negotiate the insertion of a provision into the LNG SPA which defines the circumstances in which a seller will be liable to the purchaser, and the extent and consequence of each such liability.

5.8 Pricing

Prices under LNG SPAs are typically based on price formulae linked to a reference rate, such as a crude oil price. However, as most LNG SPAs are privately negotiated and price is commercially sensitive, it is unclear whether a standard price formula has developed. The use of indexation to an external price is intended to overcome the inherent problems surrounding an LNG SPA by allowing the price to be adjusted in line with market developments. Price review mechanisms are often utilised within long-term contracts to address the same problem – the potential for the price under the contract to be out of sync with evolving market dynamics and gas price volatility.

Any price review provision must be carefully negotiated and clearly defined to avoid the mechanism becoming a means of postponing disputes, rather than acting as a proactive tool to resolve potential disputes. In order to avoid disputes and create an effective tool for price review, the price review mechanism generally defines:

- price review triggers (including event criteria, timing issues and surrounding circumstances);
- the scope of any price review (including which elements are the subject of review);

- the implementation of a price review;
- how to access the necessary information and data in a non-transparent market; and
- if the matter cannot be agreed between the parties, whether the price review is to be referred to a third-party expert or arbitrator.

5.9 Termination

An LNG SPA will end as a consequence of the natural expiry of the contract term or because an event which entitles a party to terminate under the LNG SPA has arisen and the party has exercised its right to terminate. In a supply-based LNG SPA, the term will end at the expiry of the defined term of the GSA or upon delivery by the seller of a specified gas quantity. In a depletion-based LNG SPA, the term will subsist until the gas field is depleted. In addition to the natural expiry of the LNG SPA term, the agreement will set out several events or circumstances which give rise to a termination right. Such trigger events may include:

- default of the other party;
- the seller's failure to make available or deliver gas;
- a purchaser's failure to receive gas;
- an insolvency event;
- an economic termination right in a depletion-based LNG SPA; and
- *force majeure.*

5.10 Alternative regimes

While many of the features discussed above will likely remain the mainstay of LNG SPAs, there are signs that market forces are causing some of these traditional regimes to be replaced. In an increasing number of instances, alternative regimes, often characterised by increased flexibility, are becoming increasingly prevalent.

(a) Cargo-by-cargo delivery

Whereby a buyer's failure to take is determined on a cargo-by-cargo basis (rather than an annual basis taking into account the annual contract quantity as discussed previously). Under this approach, if the buyer notifies the seller that it is unable to take delivery by a certain, pre-specified time, or if the buyer is unable to complete receipt by a pre-specified time, then the buyer is deemed to have failed to take such cargo and the buyer's obligation to compensate the seller for the resulting loss would accrue at such point in time. (This is in contrast to the traditional take-or-pay model, under which it would accrue at year-end if the total quantity taken did not equal or exceed the annual contract quantity.) This approach assists with a seller's cashflow, reduces its exposure, and mitigates the risk that payment security is less than a full year's take-or-pay obligation (leaving the seller unsecured for the balance), as it enables the seller to discontinue loadings before the contract reaches the buyer's credit support limits.

(b) Portfolio supply

Whereby a seller has the option to supply LNG from multiple sources, as opposed to supply from a dedicated project. This provides the seller with flexibility to optimise its supply and exploit short-term trade opportunities for profit (eg, favourable swap locations, shipping cost savings, arbitrage etc). This has the added advantage of providing certainty on the start date timing under the LNG SPA. Buyers who are procuring LNG to support a new LNG market or a particular project are likely to value the additional timing flexibility that portfolio sellers can offer in relation to delivery timing and contract duration. Destination flexibility on the part of a buyer is being increasingly requested in LNG SPAs, and raises its own compliance risk issues in terms of anti-bribery and corruption, sanctions and anti-trust laws.

6. Natural gas trading

The use of natural gas markets has increased dramatically in recent years, particularly with respect to LNG. The short-term LNG market has grown from virtually zero before 1990 to 16.3% of the world LNG trade in 2009 and 29% in 2015. The short-term market recognises the possibility of short-term gas price variations and provides an opportunity to profit through arbitrage. The two predominant forms of agreement utilised are short-term contracts and spot cargoes. These are supply-based contracts to deliver particular quantities of natural gas over a specified period of time at an agreed price.

Short-term or spot contracts are important risk management tools. Transparent spot markets for both wholesale gas and gas transport are key industry features in countries with developed mature gas industries, such as the United States and the United Kingdom. The experience of developed markets such as these shows that transparent spot markets that promote the efficient use of gas resources at a particular point can coexist with long-term contract markets that encourage the development of gas resources. Spot markets support and improve the efficiency of the long-term contract market by providing a ready reference for the economic value of gas while simultaneously allowing market participants to adjust the long-term positions into which they have entered. They also soften the sometimes-draconian nature of the take-or-pay clause, thus enhancing competition. Further, spot markets provide opportunities to develop some gas resources in the absence of long-term contracts.

6.1 Natural gas trading in the United States

The role of the financial contracts market for natural gas in the United States has grown as the volatility of the market has increased. Many input factors affecting price lead to profitable speculation and create an unpredictable market for players primarily concerned with the physical delivery of natural gas. Thus, those physical market players have had to respond to increased volatility with the development of short-term trading instruments for the physical supply of natural gas. These derivative supply instruments form a supplemental source of both supply and demand for natural gas producers and consumers. These short-term instruments can supplement long-term GSAs signed before shale gas came online (not uncommon, since the market transformed less than 15 years ago), or can constitute the majority of any individual stakeholder's market participation.

The most common form of contract in the natural gas markets today is a bilateral trade instrument between private players that have seats, or through brokers with seats, on the New York Mercantile Exchange (NYMEX). These instruments may be master agreements, under which trades are executed, or independent bilateral agreements. All of these short-term instruments are either fixed-price contracts, which normally have an extremely short term, or floating-index-price contracts. The US Federal Energy Regulatory Commission publishes a monthly index price based on market activity for physical supply. Since buyers and sellers meet on NYMEX, the true price reflects the physical demand for natural gas by distributors and the true availability of gas from suppliers. Natural gas producers and sellers will use this index price as the basis for a contract when they do not want to assume the risk of any of the market's volatility, and prefer to buy and sell at the true price in the physical market for that month. The Federal Energy Regulatory Commission plays an important role by aggregating the prices from all the trades that happen on NYMEX to determine the index price. Since the index price is not readily available to individual players until the data is aggregated, the Federal Energy Regulatory Commission index is used as a substitute term by parties to a contract as the price or basis for the price. Floating-index-price contracts allow a seller to ensure a profit by selling at the index rate plus a margin, or allow a buyer to ensure that it will not be paying a rate much higher than the market by agreeing to an index price plus a limited margin.

Due to the volatility of the market, floating-index-price contracts are very frequent. With the increase in volatility, the US market has seen not only a shortening of the duration of the instrument, but an overall shift in preference for these floating-price contracts. Spot trades are common and consumers of natural gas normally use these instruments to supplement or unload an existing natural gas supply. Consumers can also use hedging contracts to anticipate their natural gas needs in the future, although suppliers are increasingly hesitant to provide any long-term commitment to provide natural gas at these low prices. All of the trading is conducted from a trade sheet which reflects the prices at which natural gas as a commodity is being traded in the US market up to 2026.

6.2 Natural gas trading in Australia

The market for natural gas trading in Australia is still relatively illiquid, in that there are only a few buyers and sellers at any one time and the market does not have an accessible and managed market place or exchange. The market is contractually driven and contractual confidentiality is a cornerstone of the market, so there is little wholesale gas price transparency. However, the market has developed in recent years, as part of the Australian Government's energy market reform programme. As previously outlined, Queensland is seeing substantial developments in LNG exports, which has increased the need for more flexible and transparent upstream transactions between parties. A gas supply hub at Wallumbilla commenced in March 2014; the hub is a pipeline interconnection point for the Surat-Bowen Basin, linking gas markets in Queensland, South Australia, New South Wales and Victoria. Other gas trading markets exist in Victoria while the Short-Term Trading Market (STTM) has

Adelaide, Sydney and Brisbane operating as its trading hubs and Western Australia has a gas market. These trading markets primarily operate to manage physical imbalances and daily gas flows, in which backhaul and swaps are occasionally utilised to alter contractual supply arrangements.

(a) *Victorian spot market*

While Sydney, Adelaide and Brisbane have adopted the STTM, Victoria retains its gas wholesale market. It established a spot market for gas in 1999 to manage gas flows on the Declared Transmission System (DTS). The market allows participants to trade gas on a daily basis to address imbalances which arise from the difference between contracted gas supply quantities and actual requirements. The market framework also enables information to be gathered to facilitate pipeline operations, and maintains a reliable and secure system for gas transportation.

The market and the DTS are operated by the Australian Energy Market Operator (AEMO). A gas day lasts 24 hours and is the day before the trading day; participants submit bids to inject or withdraw gas from the DTS. At the beginning of each gas day (6am), the AEMO stacks supply offers and selects the minimum number of cost bids to match demand across the market. This establishes a spot market clearing price; gas traded at this price accounts for around 10% to 20% of wholesale volumes in Victoria. As Victoria has a net market, this price relates only to differences between contracted and actual amounts. The market has a floor of A$0 per gigajoule (GJ) and a ceiling of A$800/GJ.

As in the STTM, participants are required to make a deviation payment reflecting the difference between their actual and scheduled gas quantities. The AEMO is also empowered to schedule additional injections of gas, set above the market price, into the market to alleviate short-term imbalances. The DTS does not require participants to have gas transportation or gas supply agreements (although most participants do have underlying gas supply agreements), enabling smaller retailers to enter the market relatively easily. As the DTS uses market carriage, participants do not have the firm pipeline capacity rights of contract carriage transmission pipelines in other states. Participants can acquire an authorised maximum daily interval quantity to gain priority in times of congestion by entering into a contract with the DTS operator (APA Group).

(b) *STTM*

The STTM is designed to facilitate short-term trading using market-driven daily prices. It creates an open market, with transparent pricing that reflects realistic supply and demand, and allows long-standing arrangements between industry players to continue. As previously mentioned, the STTM has three hubs in eastern Australia: the hubs in Adelaide and Sydney were launched in September 2010, and the Brisbane hub in December 2011. All hubs are subject to the same rules and regulations, but are scheduled and settled separately.

The AEMO controls the STTM and ensures the physical supply and withdrawal of gas remains balanced. On the day before the gas day, shippers place offers to sell gas on the market and users bid to buy gas. Bids and offers can be made only by participants holding the appropriate gas rights. Once trading has closed, the AEMO determines the *ex ante* market price and provides a market schedule for the gas day. On the gas day, the pipeline operator delivers gas to the hub and users withdraw their required quantity. Gas movement that accords with the market schedule is then settled at the market price and deviations are charged based on their impact on the market. In order to maintain the physical gas balance, the AEMO provides a market operator service. Under this, arrangements are established with shippers and users who have the capacity to provide or withdraw the requisite quantity of gas. The AEMO is also empowered to authorise the use of contingency gas.

(c) Western Australia

The gas market in Western Australia differs from the eastern natural gas markets as it supplies both the domestic market and a large number of Australia's LNG exports, and is relatively illiquid. Producers value their gas products based on the market opportunity to sell to either the international or domestic market; the price of gas is therefore affected by the international LNG price and market opportunity, in addition to domestic supply and demand.

The wholesale gas market in Western Australia uses a contract carriage model. The short-term and long-term bilateral contracts that exist between producers, major users and retailers dictate supply and transportation for the majority of gas delivered into the transmission systems. Thus, there is limited potential for a secondary market to develop around auxiliary gas, and an underdeveloped infrastructure to support it. Reforms to make the market similar to the eastern market began in 2012. A gas bulletin board was implemented, providing a website for publishing information about short-term or near-term gas supply, transmission, storage and demand. It also includes emergency management instructions to assist with the management of supply disruptions. The Gas Statement of Opportunities was implemented to improve market transparency by providing an annual planning document that provides a comprehensive medium-term to long-term view of natural gas supply, demand, transmission and storage capacity.

(d) Wallumbilla Gas Supply Hub

In March 2014, the AEMO implemented a gas supply hub at Wallumbilla, Queensland. In June 2016, a trading location at Moomba (South Australia) was established, enabling participants in southern markets to trade under the same market framework and rules as Wallumbilla. The gas supply hub is a gas trading exchange through an electronic trading platform, which was introduced to enable improved wholesale trading for an east coast gas market affected by significant LNG exports in Queensland. The exchange has very little in common with the DTS or the STTM. Market participants submit orders to buy and sell short-term gas on standard terms. Each order must specify price, quantity and whether it is an offer (to sell) or a bid (to buy). The trading platform matches compatible bids and offers, with the settlement price for a transaction determined by the price in the order. Becoming a participant is voluntary (and anonymous) and participants remain free to trade bilaterally outside the exchange. To support anonymous trading, the AEMO manages prudential requirements. The AEMO centrally settles transactions and provides reports to assist participants with managing their portfolio and gas delivery obligations. The hub allows participants to trade gas over longer terms than merely daily imbalances, including weekly, monthly and three-monthly, using standardised contracts. The hub promotes transparent and efficient gas trading by allowing participants to manage the risks associated with variable gas prices. It also deepens market liquidity by attracting participants such as LNG plants, industrial customers and gas-powered energy generators. The diversity of contract positions and the number of participants at Wallumbilla creates a natural point of trade.

7. Conclusion

The global gas industry is undergoing a transformation, with the overabundance of newly economic shale gas in the United States and increased access to LNG facilities, among other things, causing a slump in world gas prices. With the commissioning of LNG export plants in the United States, Australia and elsewhere, the global gas market is expected to open up considerably. Continued exports by traditional large players – such as Qatar and Australia – and new entrants in the United States are likely to lead to excess spare capacity that will maintain prices at decreased levels into the early 2020s. This may continue the shift away from traditional long-term GSAs towards novel trading arrangements which allow players and speculators alike to take advantage of arbitraging opportunities.

The authors would like to thank Lucinda Boothby (associate, Clifford Chance) for her assistance on this chapter.

Regulatory issues related to shale gas production in the United States

Jennifer Morrissey
Dentons US LLP

1. Introduction

Nearly one quarter of the United States' total energy needs is satisfied by natural gas. Approximately two-thirds of demand comes from the industrial and electric power generation sectors, with the remaining third consisting primarily of residential and commercial uses. Gas-fired electric power generation is approaching 30% of the entire US fleet as the highly favourable economics of natural gas displace coal-fired plants.

In less than a decade, the US natural gas industry has gone from deep concern over scarcity and high market volatility, with prices exceeding $15 per million British thermal units (MMBTU), to abundant domestic natural gas supply and relatively low, stable prices in the neighbourhood of $2 to $4 per MMBTU. The change is due to the combination of new and improved directional drilling and advanced completion techniques with the longstanding industry practice of stimulating production by means of hydraulic fracturing, allowing access to natural gas resources that previously were uneconomic to extract or, in some instances, thought to be unrecoverable. Production has increased in regions with a long history of oil and gas operations (eg, in parts of Texas), but the gas boom has also affected regions less familiar with the drilling industry.

Since 2011, more than 20,000 new wells have been drilled and hydraulically fractured in the United States every year, with a number of existing wells also being fractured or refractured. Hydraulic fracturing in connection with the production of oil, gas and geothermal energy occurs in at least half of all US states. Most of this activity takes place in Texas (approximately half, by some estimates), followed by Colorado, Pennsylvania and North Dakota. Although low prices in 2016 caused production to drop off for the first time in over a decade, gas will continue to represent a significant, if not dominant, portion of the US energy fleet for the foreseeable future.

2. Overview of natural gas regulation in the United States

The US natural gas industry is open to competition across all industry segments. Natural gas production and exploration companies are privately owned and vary in size from global corporations to very small companies with one or two individuals who own only a partial interest in a single well site. While there are hundreds of natural gas production companies in the United States, a few dozen represent more than half of all domestic production.

In the United States, minerals in the ground, including oil and gas, are generally privately owned. State law governs the rights of mineral interest owners in the states, unless the United States owns minerals in that particular state, in which case federal law supersedes state law. In most states, the mineral estate can be severed from the surface estate. As such, two different individuals can own the surface and the minerals for the same tract of land. The mineral estate is the dominant estate, and the surface owner must make accommodations to allow a mineral interest holder the right to extract the underlying minerals. If minerals underlie federally owned lands, the federal government may lease the mineral rights to extraction companies in exchange for a royalty payment.

The federal and state governments derive value from natural gas production in a variety of ways, including through tax revenue, royalties, lease payments or bonuses, and permitting and filing fees. Similarly, states derive value through licensing and impact fees, and increased tax revenue related to local economic development in regions where drilling takes place.

Prices at the wellhead are not regulated, but instead are subject to market forces of supply and demand. Producers sell gas to local distribution companies, end-users or, most commonly, to marketers, which act as intermediaries and offer bundled or unbundled services and can facilitate the sale of gas, as well as contract for transportation and storage.

The United States boasts more than 300,000 miles of pipelines and hundreds of natural gas storage fields which, depending on location, are subject to federal or state regulation.

Both the federal and state governments play a role in establishing policy for the domestic natural gas and oil sectors. At the federal level, significant authority is delegated to the administrative agencies, including the Federal Energy Regulatory Commission and the Department of the Interior's Bureau of Ocean Energy Management, Regulation and Enforcement (for certain aspects of offshore facilities), the Environmental Protection Agency (EPA), the Department of Interior's Bureau of Land Management (BLM), and other federal agencies. Federal policies tend to focus on maximising the benefits of abundant domestic supplies while addressing serious environmental, health, safety and energy security concerns related to the production of natural gas.

The states also play an important role in setting policy and regulatory oversight of the natural gas sector, and with the more *laissez faire* approach of the new Trump administration, this role is likely to increase. Due to highly varied geography, as well as differing economic priorities, state-level oversight is critical to the development of the industry and to the protection of health, safety and the environment. State agencies with permitting and monitoring authority over oil and gas activity often have broad powers to adopt regulations relating to gas extraction and intrastate transportation of natural gas. These regulations typically include: permitting processes; well-construction requirements; local notice or public disclosure requirements; limits on the amount of natural gas that can be extracted from a given well; and timing and location of production.

The federal government regulates the exploration, production and transportation of natural gas in several areas. The Federal Energy Regulatory Commission maintains jurisdiction over natural gas pipelines that cross state lines (interstate pipelines), as well as the construction and operation of transportation and storage infrastructure. The Department of Transportation, through the Pipeline and Hazardous Materials Safety Administration, sets safety standards for all interstate and intrastate natural gas pipelines, as well as overseeing the transportation of crude oil by rail. Several federal agencies within the Department of the Interior manage minerals owned by the federal government and Indian tribes, regulate offshore drilling on lands owned by the US government, and oversee drilling conducted on federal onshore lands.

Policy and regulatory developments for the natural gas sector, particularly at the federal level, tend to lag behind industry developments, although this is not always evident from the national policy dialogue, which is significantly driven by increased public awareness of energy industry activities.

3. Regulation of hydraulic fracturing in the United States

There continues to be widespread public misperception in the United States that:

- hydraulic fracturing is a new technique;
- hydraulic fracturing is inherently dangerous to the environment, particularly to underground drinking water supplies; and
- the development and production of shale gas is insufficiently regulated.

This last notion finds its roots in two sources: (i) in the fact that there is no single, comprehensive national regulation of oil and gas drilling in the United States; and (ii) in a provision of the Energy Policy Act 2005, which codified a longstanding practice of excluding the disposal of most hydraulic fracturing wastewater from certain underground injection permitting requirements under the Safe Drinking Water Act that apply to the disposal of wastewater categorised under that statute as 'hazardous'.

In truth, shale gas development and production (as well as oil and conventional gas production) is highly regulated in the United States through a combination of state, local and federal regulatory schemes addressing a wide variety of issues, including:

- the impact on water from drilling and wastewater, including the injection process, wastewater treatment and gas migration, along with erosion related to site preparation;
- operational problems and accidents, such as well integrity, blowouts, spills and worker safety;
- zoning and the impact of drilling on local communities, including increased truck traffic, road damage, congestion and safety;
- air impacts of drilling operations, including methane leakage, flaring and emissions of greenhouse gases and particulate matters; and
- seismic risks.

The most targeted regulation generally occurs at the state level. Shale gas drilling takes place in many different regions across North America, each with its own geology, economic priorities, environmental concerns and experience with the oil and gas industry. Due to these and numerous other differences, in North America most agree that state regulators are far better positioned to craft and enforce appropriate rules that address the relevant issues. Federal regulation more often serves as a backstop. Nevertheless, there are several federal regulatory schemes that do apply directly to the gas industry, and there have been efforts in recent years – with varying success – to expand federal oversight.

3.1 Federal regulation

(a) The Clean Water Act

The Clean Water Act regulates surface water discharges and stormwater run-off. Surface water discharges of flowback are regulated by the National Pollutant Discharge Elimination System programme under the Clean Water Act, which requires flowback to be treated before it is discharged into surface water or injected underground. EPA rules establish pre-treatment standards for discharge of wastewater from unconventional oil and gas production.

Treatment of drilling wastewater is typically performed by wastewater treatment facilities, although there are a number of other treatment options. With respect to indirect discharges, the EPA is currently working with municipal and private wastewater treatment facilities that accept drilling waste to collect data related to the quantity of waste sent to these facilities, treatment technologies and related costs.

(b) The Safe Drinking Water Act

The Safe Drinking Water Act regulates disposal of fluid waste deep underground. Underground injection of flowback is regulated by:

- the EPA under the Safe Drinking Water Act authority under the underground injection control programme (if the flowback contains diesel fuels); or
- a state with primary underground injection control enforcement authority.

The Safe Drinking Water Act also regulates construction, operation, mechanical integrity and monitoring of wells, and requires reporting, financial assurances and public notification of injection activities.

There are five categories of underground injection well, each with its own set of rules. At present, liquid waste from hydraulic fracturing is injected into Class II wells, which are generally reserved for non-hazardous waste. There are approximately 30,000 Class II wells located across the country. Class II wells are not subject to the same rigorous testing as Class I wells (for injection of hazardous waste), although the programme includes discretionary authority that allows additional conditions to be added to the underground injection control permit on a case-by-case basis. For example, Class I wells are routinely tested for risks associated with seismicity, whereas Class II wells typically are not. However, injection into Class II wells can be conditioned on measures to reduce risk of injection-induced seismicity at the discretion of the permitting authority. Recent studies exploring the potential connection between the underground injection of waste from natural gas drilling operations and an uptick in small to moderate magnitude (less than 5.0) earthquakes in regions not generally prone to frequent seismic activity have prompted discussions about either recategorising hydraulic fracturing waste as 'hazardous' or modifying the underground injection control rules to require additional testing of Class II injection wells. The latter step is more likely, since in most cases it will be difficult to demonstrate that flowback from hydraulic fracturing is significantly distinguishable from other drilling waste that falls into Category II. In 2015, at the request of the EPA, a technical working group issued a report making recommendations for managing and minimising the potential for significant injection-induced seismic events (ie, earthquakes of a magnitude that could potentially cause damage to or endanger underground sources of drinking water) using the existing Class II regulatory framework.

It is also noteworthy that the EPA has guidelines issued under the Safe Drinking Water Act for federal permitting requirements for hydraulic fracturing when diesel is used as a component of the fracturing fluids. Unlike most hydraulic fracturing activity, drilling with diesel as a component of the fluids that are pumped into a well has never been exempt from the Safe Drinking Water Act, and a separate permit under the underground injection control rules has consistently been required. Although diesel use is less widespread today than when the Safe Drinking Water Act was initially enacted, it is still used where no good substitute is available (eg, when drilling in very cold climates).

The EPA's guidance applies specifically to federal agencies that permit underground injection of drilling waste that includes diesel fuels; however, the guidance incorporates many practices that are consistent with state regulations and widely used voluntary standards from the American Petroleum Institute, among other guidelines.

(c) The Toxic Substances Control Act

In November 2011 and again in Autumn 2012, the EPA announced that it would begin exploring the possibility of initiating a rulemaking under the Toxic Substances Control Act on the disclosure of chemicals used in fluids in the hydraulic fracturing process and the identification of health and environmental risks associated with their use. As many states already have disclosure requirements (at least 20), it was unclear how any such federal rules, if implemented, would interact with state rules. The public comment period on the first phase of this initiative closed in 2014, but the EPA did not begin a formal rulemaking process. Then in 2015, the EPA rejected a petition filed by environmental groups to issue new controls for mercury and carbon dioxide under the Toxic Substances Control Act, finding that existing controls are sufficient to address public health and environmental risks.

(d) Natural gas drilling on US federal and tribal lands

In 2015, the BLM issued rules relating to the protection of water quality in connection with hydraulic fracturing activities on federal and tribal lands. While the impact of these rules would affect drilling activity on public lands across the United States, the brunt is directed at western states where far more land is under federal authority. The rules are deferential to many existing state laws and provide a fair amount of flexibility to industry. Among other things, the rules require disclosure of chemical ingredients in fracturing fluids and establish requirements for wellbore construction and integrity, and for wastewater management. Variances may be available to lands within the boundaries of states where the agency determines that existing state regulations meet or exceed the standards set forth in the federal rule. Disclosure of chemicals used in the fracturing process may be carried out through the widely used FracFocus website and, similar to some existing state rules, trade secrets may be protected.

In Summer 2016, a federal district court judge in Wyoming ruled that the BLM had exceeded its statutory authority in promulgating the regulations. The BLM asserted that it had broad authority to implement the rules under several federal statutes taken together, including statutes addressing federal leasing of mineral rights, land management and Indian mineral development and leasing. The court examined the authority delegated to the BLM by Congress under each of those laws and concluded not only that the laws in question did not authorise the BLM to regulate hydraulic fracturing in the manner that it did, but also that Congress has expressly removed hydraulic fracturing from federal regulation in the Safe Drinking Water Act (see previous discussion). The court held that the fact that the Safe Drinking Water Act limits the EPA's authority to regulate hydraulic fracturing does not mean that other agencies are free to regulate. To the contrary, the EPA has authority to regulate all underground injection and Congress' express carve out of hydraulic fracturing from this federal regulatory scheme is definitive. States and Indian tribes may assume primary enforcement responsibility, subject to EPA approval. Congress left no room for concurrent regulation by the BLM.

At time of publication, this case is pending decision by a federal court of appeals.[1]

(e) Clean Air Act rules applicable to the natural gas industry

In August 2012, the EPA published a regulation requiring "green completions" for all new or refractured gas wells after 2015. Operators are required to capture air emissions (methane and volatile organic compounds) using reduced emission completion technology. Existing wells or wells that predated the rule were required to use either reduced emission completions or combustion control devices. The EPA's rules also include notification and record-keeping requirements.

In May 2016, the EPA released a final rule that establishes Clean New Source Performance Standards for methane emissions from various new, reconstructed and modified oil and gas equipment and processes. At the same time, the agency initiated an Information Collection Request to gather information relating to emissions from existing sources in the oil and gas industry. It was expected to lead to rules on existing sources, but with the change of administration in January 2017, the future of these rules is uncertain. In 2012, the EPA had issued rules addressing other pollutants relating to oil and gas production and processing, but those rules did not directly target methane.

---

1 *Wyoming v Jewell*, Case No 16-8068 (10th Circuit).

(f) Other potentially applicable federal rules

The National Environmental Policy Act: The National Environmental Policy Act requires permits and environmental impact assessments to be conducted before drilling on federal lands. The potential impact of natural gas drilling and production on water resources within the area under review may have an impact on the conclusions drawn and the restrictions or conditions imposed on the proposed drilling activity.

The Occupational Safety and Health Act: The Occupational Safety and Health Act sets standards to keep workers safe. These rules include requiring companies to maintain material safety data sheets for chemicals used by workers and to make such data sheets available onsite. State regulators also frequently require data sheets to be filed and made available to the public in connection with the issuance of various permits. The act has issued a rule on airborne crystalline silica exposure and has specifically highlighted the need to protect workers involved in oil and gas drilling activities (among other industries) from silica dust.

Emergency Planning and Community Right to Know Act: The Emergency Planning and Community Right to Know Act requires storage of regulated chemicals above certain quantities to be reported annually to local and state emergency responders.

The Securities and Exchange Commission: The Securities and Exchange Commission has also inserted itself into the hydraulic fracturing conversation, sending letters to drilling companies seeking information about chemical use, including quantification estimates, to ensure accurate public disclosure.

Comprehensive Environmental Response, Compensation and Liability Act: The EPA took action pursuant to its authority under the Comprehensive Environmental Response, Compensation and Liability Act 1980 (also known as Superfund) to conduct a groundwater investigation of the Wind River water formation near the town of Pavillion, Wyoming. The shale plays in this area are very shallow and elevated levels of methane and diesel were detected in a handful of wells in the town. The EPA reported that more study was required, but that initial results of testing indicated that hydraulic fracturing may have contributed to groundwater contamination.[2] The report was widely criticised, including by the US Geological Survey, because the investigators lacked baseline data against which to measure contaminants found in the wells.[3]

---

2 Dominic C DiGiulio, Environmental Protection Agency, "Investigation of Ground Water Contamination Near Pavillion, Wyoming" (2011), available at www.epa.gov/region8/superfund/wy/pavillion/EPA_ReportOnPavillion_Dec-8- 2011.pdf

3 See, for example Peter R Wright, "Groundwater-Quality and Quality-Control Data for Two Monitoring Wells near Pavillion, Wyoming", April and May 2012, US Geological Survey (2012), http://pubs.usgs.gov/ds/718/.

3.2 State regulation

At the state level, there are nearly as many approaches to the regulation of hydraulic fracturing (and of natural gas drilling and production generally) as there are US states in which these activities take place. The US system allocates responsibility to both federal and state authorities, putting regulation in the hands of the entity that is best placed to deal with a particular issue. As a general rule, where federal regulations apply, a state's regulation must be at least as protective as the federal standard. Differing geology and local priorities dictate differing regulation. Each state in which the gas industry is active, or where drilling may occur in future, is in the process of shaping its own regulatory and permitting scheme to meet the needs and priorities of its own population.

In general, the western states have lengthier experience with drilling and therefore have more mature – and less controversial (at times) – regulatory systems. However, this is not to say that oil and gas drilling is widely accepted in all western localities. In recent years, there have been several attempts to enact local hydraulic fracturing bans in Colorado, Texas and California, with varying degrees of success. The experience of the eastern states with natural gas drilling is more recent, although many states have well-developed laws on mining and Pennsylvania has a long history of oil drilling. The states in the Marcellus region in the northeast found themselves having to respond quickly to a rapidly developing industry. Natural gas producers, accustomed to operations in the western and Gulf Coast region, have had to adapt to regulatory schemes that view their activities with mistrust and suspicion.

Pennsylvania, the eastern state with the most active natural gas industry, has a vigilant and engaged Department of Environmental Protection. Driven in part by the economic opportunities that shale gas presents, the state has moved quickly to adapt its laws to accommodate drilling companies, while striving to protect communities and water resources. A number of years ago, the state enacted comprehensive legislation that includes an impact fee to be provided to localities to address local infrastructure concerns that arise as towns experience sudden growth prompted by an influx of gas companies. Many of the rules were challenged in court, and the state recently adopted new rules that appear to be more widely accepted by environmental groups and affected communities. State regulators have taken a proactive approach to identifying gaps in the law and to closing those gaps, but in a thoughtful and measured manner based on data and interaction with the companies operating within the state.

Several states have enacted bans or moratoriums on hydraulic fracturing. In 2015, New York banned high-volume hydraulic fracturing associated with horizontal drilling techniques after a multi-year moratorium was implemented to give various state agencies time to study the associated environmental and health impacts. Other states in the region have enacted bans (eg, Vermont) or temporary moratoriums while they watch to see how Pennsylvania's experience plays out and while they develop their own bodies of regulation (eg, Maryland and New Jersey). In New Jersey, Pennsylvania, Colorado, Texas and California, towns and municipalities have also passed ordinances temporarily or permanently banning hydraulic fracturing within town limits. In a number of instances, these local bans have been struck down by the courts on the basis of pre-emption of state law.

Even in regions with lengthy experience of oil and gas drilling, public concerns have led to the enactment of specific regulations for hydraulic fracturing. California, Illinois and Ohio have recently passed new rules applicable to oil and gas operations, and other states, such as Michigan and North Carolina, are in the process of drafting rules. Many of these rules have to do with disclosure of chemicals used in fracturing processes, which remains an issue of primary concern to the public. To some extent, drilling occurs in more than 30 US states, and what is required in terms of disclosure varies widely. Disclosure of chemicals used in the drilling and well stimulation process is sometimes required to be made only to state authorities, although more and more companies are voluntarily disclosing the contents of fracturing fluids on the FracFocus website. More than 117,000 well sites are currently included in the FracFocus database.

Regulations also typically include rules on well construction, testing and monitoring, treatment and disposal of fracturing waste. Several states, such as Colorado, Wyoming and Illinois, have developed or are developing laws to curb emissions, including requiring green well completions to capture methane and emissions, or requiring drilling companies to conduct regular testing. Other states are likely to follow suit.

In addition, a number of states require drilling companies to provide notice to local landowners when hydraulic fracturing occurs.

3.3 Seismic concerns

In the past couple of years, certain regions of the country where enhanced oil and gas production takes place have experienced a significant uptick in seismic activity. Oklahoma is most often in the headlines, but Texas, Ohio, Alabama, Arkansas, Colorado, Kansas and New Mexico have also experienced an increase in earthquakes. Initially, speculation was that hydraulic fracturing might have caused the increase; however, recent studies appear to indicate that it is not the enhanced recovery but rather the injection and storage of wastewater from drilling activity (conventional oil and gas drilling and hydraulic fracturing) that may be to blame. Several studies are under way, including by the US Geological Survey and a number of universities, but research appears to indicate that in regions with active faults, high injection rates (eg, several hundred thousand barrels per month) may increase the risk of earthquakes, although other factors are also in play, such as: well-head pressure; the total volume of fluid injected; the rate of injection; the number and proximity of disposal wells in a region; and the geological formation into which the wastewater is injected. The high number of recent earthquakes in certain regions of Oklahoma has prompted local regulators to close storage wells in some parts of the state until the link, and possible mitigation, can be more fully understood. Other regions in which drilling has increased significantly in recent years (eg, North Dakota) have not recorded any increased seismic activity linked to drilling activity, likely due to different subterranean geology.

Underground injection of drilling waste is a lucrative business for some states, such as Ohio, which has numerous disposal wells and fairly lax disclosure requirements compared to other states, and is adjacent to the Marcellus region.

Pennsylvania, the hub of Marcellus activity, has over 1,800 Class II injection wells, but only a small handful are approved for oil and gas waste disposal.

3.4 Overview of regulation of water use and water quality in hydraulic fracturing

Water use and quality are the most commonly raised concerns relating to hydraulic fracturing, so it is unsurprising that regulators are giving significant attention to potential impacts of hydraulic fracturing on surface and groundwater supplies.

Water is an essential tool in the production of natural gas. In addition to the water used in the drilling process to construct the well, hydraulic fracturing techniques used to stimulate the well require up to 5 million gallons of water for each well. The fluids, including produced brines, that are recovered from the well – known as flowback and 'produced water' – require proper treatment and disposal. Moreover, the drilling process itself can often involve boring a hole through the water table to reach the shale plays thousands of feet below where the gas is trapped. This process has given rise to public concerns centring on:

- the protection of underground sources of drinking water from drilling and well stimulation activities;
- wastewater treatment and disposal; and
- the regulation of water use, particularly in times of scarcity.

(a) Regulatory issues related to water quality

The key concern raised by environmentalists, as well as by the public in the communities in which natural gas drilling takes place or is planned, is the potential for contamination of underground and surface water sources during drilling or hydraulic fracturing. Their focus has been on the possibility of gas or contaminants from a fractured well migrating through layers of rock into the drinking water supply thousands of feet above. While theoretically possible, research and practical experience suggest that the real risk to water quality is the potential contamination of surface and groundwater resources from run-off and erosion accelerated by drilling-related activity, by spills of contaminated flowback, and by inadequate methods for the treatment and disposal of drilling wastewater.

Another real risk to water quality is a leak into groundwater sources due to faulty cement casings where the well passes through the water table. Wellbore integrity is generally regulated by the states and is increasingly the subject of self-policing, as many industry players attempt to devise best practices. A number of years ago this issue attracted significant media attention in connection with well-water contamination in Dimock, Pennsylvania. The incident was blamed in part on faulty well casings, although the source of contamination was never conclusively identified.

Concerns about wellbore integrity and the possibility of migration of contaminants are driving demands for greater industry disclosure regarding the fluids that are pumped into a well to fracture the shale and release the gas trapped within the formations. The fracturing fluids are approximately 99.5% water, but when millions of gallons are at issue, the chemicals that make up the remaining 0.5%, together with the produced brines, are frequently viewed as a sufficient quantity to merit concern. Even where disclosure requirements apply, other issues may arise, such as perceived loopholes in existing regulatory schemes to the ability of state environmental regulators in gas-boom regions to handle the ever-increasing number of environmental permit requests.

The management and disposal of drilling wastewater are also being more closely scrutinised by regulators and the public. Recycling of fracturing wastewater is becoming an industry standard, but drilling liquids cannot be recycled indefinitely. Eventually, the waste produced must be treated and disposed of. After multiple uses, concentrated brine has extremely high salt levels, which complicates disposal. Efforts are underway to find uses for brines – for example, some states use it for street de-icing in winter.

Storage of waste fluids in containment pits carries a risk of leakage or spills, particularly in the northeastern states where frequent severe weather may affect the integrity of the pit. Although this is less of a problem in the dryer states of the west, the land application is not always an option for large volumes of wastewater. Even where it is feasible, great care is required to avoid the harmful effects of overloading an area with wastewater.

Wastewater treatment facilities are not always equipped to handle drilling waste, leading to fears that inadequately treated wastewater is being released into streams. New Jersey, which has no active natural gas drilling industry, has taken steps to ban the processing of hydraulic fracturing waste in the state out of fear that its waste treatment facilities may become an attractive disposal alternative for companies in the Marcellus.

Some companies are experimenting with replacing some of the more commonly used chemicals with relatively harmless ingredients in their fracturing fluids – including substances also used in food additives. However, there is no industry standard and each company's specific 'recipe' (if not the ingredients) is a closely guarded secret.

In addition to the chemicals used in fracturing fluids, concerns are raised about wastewater levels of total dissolved solids – that is, the total of all organic and inorganic substances in a dissolved state in water, mostly inorganic salts. Levels of total dissolved solids affect the health of a stream; high levels of chlorides and salts may disrupt its system balance. Although not generally a threat to human health or aquatic life, this affects the odour and taste of water, which can lead to complaints of contamination. The most common sources of such dissolved solids are mining water discharges and urban stormwater run-off, but they are increasingly being linked to natural gas drilling.

The underground injection of drilling waste, which was long considered the preferred disposal method for liquid drilling waste, has recently been called into question because of suspicions that disposal into deep injection wells may trigger earthquakes (see previous discussion). This issue is not unique to the disposal of fracturing fluids, but because of greatly increased volumes of drilling waste generated by the shale boom, the spotlight on the natural gas industry is inescapable.

Variations in geology across the country pose further issues. A few years ago the image of a resident in a town with active gas drilling setting his tap water on fire created an internet sensation. The implication was that drilling activity had contaminated his well water. In many regions of the country, shallow methane underground has been known to cause significant concentrations of methane to collect in wells, irrespective of the proximity of drilling. When this is combined with lax permitting requirements for the construction of drinking water wells, the result is flammable water. In most regions, the lack of baseline data for drinking water quality makes it difficult to assess the connection, if any, between gas production and high levels of methane or other contaminants in drinking water supplies.

(b) Studies on the link between water contamination and natural gas drilling

Over the past several years, numerous studies have been commissioned to evaluate the potential risks of contamination to US water supplies posed by hydraulic fracturing. Some of these initiatives are directed by federal and state regulatory agencies, while others are funded by the industry or by environmental groups. Unsurprisingly, the various industry and interest group studies tend to support the views of the groups funding them, and are viewed as reports by 'duelling experts', some of which have acquired greater credibility than others.

In December 2016, the EPA issued a report on the impacts of the hydraulic fracturing water cycle on US drinking water resources.[4] About a decade ago, the EPA looked at the potential impact on drinking water supplies of hydraulic fracturing in the production of coalbed methane and concluded that the practice posed little, if any, threat to underground drinking sources.[5] Beginning in 2011, the EPA was asked to expand the study to examine the potential impact of hydraulic fracturing in much deeper shale gas wells. In this report, the EPA looked extensively at the potential impacts on drinking water resources of water withdrawals, chemical mixing, well injection, flowback and produced water, and wastewater treatment and disposal. The EPA used a very broad definition of drinking water resources including ground and surface waters, fresh and non-fresh, which are or could potentially be used as drinking water. (Most federal and state regulatory definitions are much narrower in scope.) In a preliminary draft of the report issued in 2015, the EPA concluded that it "did not find evidence that these mechanisms have led to widespread, systemic impacts on drinking water resources in the United States."[6] The EPA did note in that draft specific instances where there have been localised impacts on drinking water, including contamination of drinking water wells, but found that the "number of identified cases... was small compared to the number of hydraulically fractured wells."[7]

---

4 "Hydraulic Fracturing for Oil and Gas: Impacts from the Hydraulic Fracturing Water Cycle on Drinking Water Resources in the United States", US Environmental Protection Agency, EPA-600-R-16-236ES (Dec 2016), www.epa.gov/hfstudy.

5 "Evaluation of Impacts to Underground Sources of Drinking Water by Hydraulic Fracturing of Coalbed Methane Reservoirs Study", US Environmental Protection Agency, available at https://fracfocus.org/sites/default/files/publications/evaluation_of_impacts_to_underground_sources_of_drinking_water_by_hydraulic_fracturing_of_coalbed_methane_reservoirs.pdf (visited 12 May 2016).

6 "Assessment of the Potential Impacts of Hydraulic Fracturing for Oil and Gas on Drinking Water Resources" (External Review Draft). US Environmental Protection Agency, Washington, DC, EPA/600/R-15/047, 2015, available at https://www.epa.gov/hfstudy/executive-summary-hydraulic-fracturing-study-draft-assessment-2015

7 *Id.*

In August 2016, the Science Advisory Board released an assessment of the EPA's report. The board did not take issue with the EPA's approach or the validity of the overall conclusions, but did suggest that for the sake of clarity, the agency provide additional support for some of the major findings. The board was concerned that the EPA's national level conclusions may minimise localised impacts. In other words, the EPA's characterisation that there is no "systemic" and "widespread" impact may not be fully understood by the public in the context that the EPA intended, and could set the agency up for criticism. Among other things, the board urges the EPA in its final report to include more discussion of the toxicity of the injected fluids and the produced waters, including whether constituents of those fluids are somehow unique to the hydraulic fracturing process (as opposed to other oil and gas production methods), and whether they are naturally occurring in the formation that is being fractured.

The board also suggested that more baseline data would be helpful. A lack of baseline data about the conditions of water sources in proximity to hydraulic fracturing activity has been a source of frustration to a number of state and local regulators as well, as it is difficult to evaluate the impact of oil and gas drilling activity on local underground drinking water supplies, and on streams where drilling waste might be discharged, without good data. In some regions, particularly where there is shallow coalbed methane, methane and other contamination in water supplies pre-dated drilling activity. Similarly, in some regions where there is lax regulation of water well construction, gas and other contaminants sometimes end up commingling with the well water.

In the final report issued in December 2016, the EPA adopted many of the board's suggestions. The final report has moved away from the "no systemic or widespread impact" and instead focuses on potential localised impacts at each stage of fracturing operations. Decision-makers relying on the report are urged to direct their attention to local or regional issues in order to reduce vulnerabilities to drinking water resources, but to keep in mind that the technologies used by the oil and gas industry are rapidly evolving, and studies and regulations will have to keep pace with this evolution.

(c) Regulation of water use – water use in context

The quantity of water required in a typical hydraulic fracturing job, when viewed in isolation, may seem quite large. However, it represents only a fraction of municipal and industrial water uses. In fact, shale gas is one of the least water-intensive energy sources to produce. To place it in context, a typical fracturing job uses between 2.5 and 5 million gallons of water. Approximately 4 million gallons of water are consumed in New York City every six minutes, and the same amount is typically used to water the average golf course every three weeks. Electric power generation, agriculture, industry and municipalities are by far the largest water consumers, accounting for 90% of all water withdrawals in the United States. According to the US Geological Survey, approximately 161,000 million gallons a day (mgd) are used in thermoelectric power generation in the United States – the scrubbers on a single coal-fired plant may use up to 5 mgd. Nationwide, 120,000 mgd are used for agriculture; 42,000 mgd for the public water supply; approximately 16,000 mgd for industry; and about 5,300 mgd for mining activities, including oil and gas extraction.[8] All mining activity nationwide – including mineral, oil and gas extraction – accounts for approximately 1% of total water withdrawals from surface and groundwater sources,[9] and although regional percentages vary, it is consistently a very small percentage of total use. In terms of the water used for each unit of energy produced, shale gas is low on the scale. At 0.5 to 3 gallons per MMBTU, it uses far less water than coal (up to 32 gallons per MMBTU), nuclear (8 to 14 gallons per MMBTU), oil (8 to 20 gallons per MMBTU) or agricultural biofuels, such as corn ethanol or soy biodiesel (more than 2,500 gallons per MMBTU).

The water used in hydraulic fracturing comes from a number of sources. Most is purchased from municipal water suppliers, although some is withdrawn from surface water. Very little comes from groundwater sources. Some companies obtain their water supply from treated wastewater from municipal and industrial treatment facilities or cooling water from power plants; increasingly, recycled produced water or flowback water is used. A few companies are also working to develop methods to locate and use sour-water aquifers.

Water use in connection with hydraulic fracturing is mostly considered 'consumptive'. Although some flowback can be treated and released into water sources, it is rarely returned to the basin of origin. Furthermore, a substantial quantity of water will not return to the water cycle, but will remain in the ground, either in the gas well or in a disposal well.

The importance of the issue of water use partly depends on the region of the country in which drilling takes place and the competing uses for the water in that region. In areas where water is scarce, withdrawals from local sources may be viewed as significant.

(d) Federal and state legal regimes for water use

Water withdrawals are primarily a matter of state law. The federal government may become involved to the extent that water withdrawals are made from waters on federal lands or from federal multi-purpose water projects, such as federally managed reservoirs, or to the extent that they affect navigable streams under federal jurisdiction. The federal government may also become involved where:

- tribal water resources are at issue;
- there are interstate disputes over rights to waters flowing through streams; or
- fish or other wildlife species are at risk.

Various federal agencies may be involved, depending on the particular water resource; however, most regulation of water use occurs at the state or regional level.

---

8 See MA Maupin, JF Kenny, SS Hutson, JK Lovelace, NL Barber and KS Linsey, 2014, "Estimated use of water in the United States in 2010", US Geological Survey Circular 1405, 56 p, https://dx.doi.org/10.3133/cir1405, also at https://pubs.usgs.gov/fs/2014/3109/pdf/fs2014-3109.pdf and https://water.usgs.gov/watuse/. Statistics are for 2010, the last year in which this data was collected. The report is published every five years. Data for the 2015 report is still under analysis as of publication of this chapter.

9 *Id*. Nearly 60% of water withdrawals for mining activities is saline water, not freshwater.

There are three primary water use regimes in the United States. Most states east of the Mississippi River follow a common law system of riparian rights, whereas most western states apply a regime based on prior appropriation or first beneficial use. Some states follow a hybrid (or 'regulated riparian') system.

Under the riparian rights rule, a water resource is shared by riparian users (ie, all those whose property touches the water resource). There is no right to a certain amount of water during a shortage or when various uses are in competition. Generally, the upstream landowner has priority with respect to water in streams that touch the property and may withdraw unlimited amounts of water for domestic and reasonable extraordinary use on the property. The landowner may withdraw groundwater underlying the property, regardless of the effect on neighbouring tracts. Downstream landowners have no security of water rights.[10] Traditionally, a transfer of water resources off the property was considered unreasonable *per se*, but this view has been modified to allow for some non-riparian uses in some regions, including municipal water supply.

The rule of prior appropriation is a 'first in time, first in line' rule. For surface waters, the original user may generally take as much water as is needed, regardless of whether the user is downstream of subsequent competing uses. There is no obligation to share the resource. Where two users are equal in their position in line, priority goes to the 'higher' use. Agriculture and irrigation are generally considered the highest uses.

Groundwater is treated differently from one state to another. In some states, ownership of groundwater is treated as a property right, similar to ownership of minerals or oil, and the resource is deemed to be 'owned in place' – meaning that the landowner has the right to capture and pump as much as is available, regardless of the effect on neighbouring landowners. In some states, groundwater remains with the surface estate if mineral rights are severed; in others, it may go with the mineral estate.

In a regulated riparian system, water rights are allocated for periods of time by means of a permit with conditions, such as minimum flow requirements, to protect downstream users. Restrictions on transfers are eased, but there may be a requirement to compensate the basin of origin. Approximately 18 states follow a hybrid water rights regime. The mid-Atlantic region of the United States, which generally corresponds to the region overlying the Marcellus Shale, has regional regulatory bodies in addition to the state regulators overseeing water quantity and quality issues under a regulated riparian system. The Delaware River Basin Commission comprises representatives from New Jersey, New York, Pennsylvania, Delaware and the US Army Corps of Engineers, while the states of Maryland, Pennsylvania and New York are represented on the Susquehanna River Basin Commission. Each commission provides drinking water to millions of customers within its region and is responsible for handling permit requests for water withdrawals in its river basin. These commissions have the authority to require

10 This is the 'American Rule' and may not apply in non-US jurisdictions that follow a rule of riparian rights.

reductions in consumptive uses during droughts and can require payment of fees to provide storage in reservoirs for release during low flows. In addition, the commissions monitor water quality, resolve disputes over competing uses, and issue exploration and drilling permits within their respective jurisdictions.

Shale and oil disputes

Scot Anderson
Hogan Lovells
Michael Darowski
Gowling WLG
Andrew Lillie
Elizabeth Titus
Hogan Lovells

1. Introduction

The unconventional development of shale formations and tight sands for oil and gas production is truly revolutionary. Any revolution, however, creates turmoil, and the shale revolution has generated its share of legal disputes, many unique to shale oil and gas development. Since most shale development to date has occurred in the United States, this chapter focuses primarily on legal disputes in the United States, before concluding by briefly considering the position in Europe – the likely next frontier for shale gas development.

2. The United States

2.1 Introduction

(a) *The nature of shale development: driver of a US energy revolution*

Over long periods of time, dead vegetation and sediment (rocks, soil) in lake and river bottoms become highly compressed, forming underground shale deposits.[1] Most shale deposits contain fractures and fissures that trap oil and natural gas.[2] Similarly, over time the sands on the shores of such lakes and rivers became sandstone, now characterised by petroleum geologists as 'tight sand formations', which can also hold oil and gas.[3]

Both shale and tight sand formations are characterised by low permeability and, accordingly, such formations have not historically been amenable to conventional oil and gas production. To access the 'trapped' oil and gas, producers developed unconventional strategies. Specifically, they refined the use of hydraulic fracturing and the engineering of long horizontal wells; techniques that were first developed many decades ago but were unnecessary for conventional well development. Modernisation, deployment and refinement of these techniques, and the use of them in combination over the past 20 years, finally unlocked oil and gas historically trapped in the fractures, fissures and tight sands. This unleashed a new era in US energy production.

---

1 David Summers, "The Differences in Fracking Tight Sand and Shales" (6 August 2014), available at http://oilprice.com/Energy/Natural-Gas/The-Differences-In-Fracking-Tight-Sand-And-Shales.html (last accessed 19 December 2016).
2 *Id.*
3 *Id.*

Hydraulic fracturing, popularly referred to as 'fracking', is a technique drillers and well-completion specialists use to stimulate oil and gas production – in short, to free the oil and gas otherwise trapped in the subterranean formations. After the wellbore is drilled and down-well portions of the well are perforated to create access to the surrounding shale or tight sands, the operator injects water and a solution of chemicals into the formation. The chemicals lubricate the well and encourage the retrieval of product down the wellbore, all the while under tremendous amounts of pressure. The water solution finds its way into the rock formation, fracturing it and creating greater permeability, which promotes the movement of underground hydrocarbons to the wellbore. After the fractures form, the operator pushes what is called 'proppant' down the wellbore – usually sand, but sometimes manufactured particles resembling sand. These small particles flow into and occupy the fractured space created by the pressurised water solution, holding the fractures open and extending the beneficial life of the well. The operator then flushes the water from the well, recycling or ultimately disposing of it.

Horizontal drilling is also critical to shale and tight-sands development. A classically conceived oil or gas well is vertical, drilled straight through the ground to reach an underground trap or structure loaded with hydrocarbons. To drill a horizontal well, the operator instead drills a well vertically to the target formation and then turns the well horizontally to traverse that formation. Doing so permits access to vastly greater volumes of hydrocarbons over much larger underground areas. In addition, horizontal drilling significantly reduces the surface-level effects of operation and production because it is common to drill and maintain numerous horizontal wells from one well pad, rather than using one pad for each well. Indeed, wells of 10,000 feet in length have become routine.

Oil and gas operators have been using hydraulic fracturing and horizontal drilling for decades. However, by 2000, Mitchell Energy had perfected these processes in combination, experiencing great success in the Barnett Shale.[4] In 2000, the United States produced just over two billion cubic feet (bcf) of natural gas each day; today the rate of production is well over 40 bcf per day. It is no exaggeration to say that shale development – or what has become known as 'unconventional' oil and gas development – has created a revolution in US energy exploration and production, as well as newfound energy independence for the country.

(b) Controversies

Every revolution has its consequences – intended and unintended – and the legal issues associated with shale development are no exception. Indeed, the Texas Court of Appeals recognised that oil and gas law may need to diverge from principles developed in a world of conventional, vertical wells:

> *[W]e recognize the immense benefits that have accompanied the advent of horizontal drilling, including the reduction of waste and the more efficient recovery of hydrocarbons... We decline to apply legal principles appropriate to vertical wells that are*

4 "Shale in the United States", Energy Information Agency, available at https://www.eia.gov/energy_in_brief/article/shale_in_the_united_states.cfm (last accessed 19 December 2016).

> *so blatantly inappropriate to horizontal wells and [that] would discourage the use of this promising technology.*[5]

Since shale development creates smaller surface footprints due to an operator's ability to drill multiple long horizontal wells from a single well pad, and to use centralised processing facilities for that production, this form of development, in theory, should create fewer land-use conflicts. However, shale development, which usually depends on hydraulic fracturing, still gives rise to many varied legal claims focused on, among other things:

- land-use issues;
- environmental issues (including water use, water contamination and air pollution, whether factually substantiated or not); and
- associated human-health questions.

These issues generally shape the terrain for disputes, but the fact that so much shale development occurs in close proximity to residential areas compounds these concerns and has led to unprecedented controversy for the oil and gas industry.

In the past, oil and gas production occurred in remote areas – on far-flung ranchland or back pastures, on Bureau of Land Management land far from towns and cities, or a couple of hills over from someone's farm. Now, growing human populations, coupled with an operator's ability to develop resources that were previously uneconomic, has led to dramatically increased interaction – and hence friction – between oil and gas operations and the people who live near them. As Smith and Richards note:

> *[m]any shale resources that were previously considered unrecoverable are now economically recoverable. Interestingly enough, a significant portion of these shale resources encroach upon urban developments. As many as 300 million people around the world across six continents occupy land overlying a shale reservoir. Large-scale industrial extraction of shale will no doubt impact these urban developments.*[6]

These observations could not be more on point, and the impacts associated with them create legal problems. The remainder of this chapter discusses the disputes that arise from the deployment of unconventional techniques required to bring about and sustain the shale revolution.

2.2 Split estates

Oil and gas reserves exist underground and require access from the surface to become marketable. Under federal law, the minerals in the ground (the mineral estate) might be owned by someone other than the owner of the surface (the surface estate). The traditional rule is that the owner of land owns the area from the surface to the centre of the earth, and then upward into the sky. However, that rule slips away for underground oil and gas resources:

---

5 *Browning Oil Co v Luecke*, 38 SW 3d 625, 642, 647 (Tex App Austin 2000, pet denied), as discussed in Bret Wells, "The Dominant Mineral Estate in the Horizontal Well Context: Time to Extend *Moser* Horizontally", 53 *Houston Law Review* 193 (2015).

6 D Smith and J Richards, "Social License to Operate: Hydraulic Fracturing-Related Challenges Facing the Oil & Gas Industry", Rocky Mountain Mineral Law Foundation, Special Institute on International Mining and Oil & Gas Law, Development and Investment (April 2015).

When you own property in the West you don't always own everything from the surface to the center of the Earth. Someone else might own the minerals lying underground and the right to access them. Someone else still might own the right to use the water flowing through your property. All this can invite confusion – and litigation.[7]

In the United States, it is common for the mineral estate and surface estate to be split. A private landowner might sell his mineral rights to a mining company, retaining his surface interest. In the settlement of the western United States, the federal government would often issue a patent for surface lands, while retaining some or all of the mineral interests.[8] When the United States brought new states into the Union, it would grant sections of land to the new state, which would be reserved for state governmental purposes. The creation of Indian reservations, and the subsequent diminishment of those reservations through settlement, allotments and withdrawals, also created split estates. Thus, a parcel of land might contain tribal minerals and the rancher might own the surface in fee, or a parcel might contain federal minerals and a state-owned surface. Many other permutations of the split estate exist.

Disputes over the development of shale often arise when a private party (or parties) owns the surface and another entity owns the minerals. The mineral owner might be an oil and gas company, but, more often, the owner of the minerals (a private party or a governmental entity) leases the minerals to the oil and gas company for energy development.

Traditionally, the law holds that a mineral estate is dominant, favouring the underlying severed mineral owner over the surface owner.[9] More recently, however, courts and some legislatures have adopted an accommodation doctrine in which the needs of both owners are balanced.[10] In Colorado, for example, the Colorado Supreme Court in *Gerrity v Magness*[11] held that a mineral-estate owner is "privileged to access the surface and use that portion of the surface estate as is reasonably necessary to develop the severed mineral interest." The Colorado legislature later codified the accommodation doctrine as announced in *Gerrity* in statute: "An operator shall conduct oil and gas operations in a manner that accommodates the surface owner by minimizing intrusion upon and damage to the surface of the land."[12]

Although the accommodation doctrine should, in theory, allow for cooperative behaviour between the owner of the surface and the owner (or the lessee) of the mineral estate, disputes still arise. A surface owner may, for example, allege that the operator's use of the surface did not minimise intrusion or damage to the surface. Under the common-law formulation of the accommodation test, a surface owner

---

7 *Entek GRB LLC v Stull Ranches LLC*, 763 F 3d 1252 (10th Cir 2014), cert denied, 135 S Ct 1895 (2015).

8 See for example, Stock-Raising Homestead Act 1916, 43 USC Sections 299, 301.

9 See for example, *Mingo Oil Producers v Kamp Cattle Co*, 776 P 2d 736 (Wyo 1989); see generally Rick D Davis, Jr, "Private Lands – Surface Access and Use", *Severed Minerals, Split Estates, Rights of Access, and Surface Use in Mineral Extraction Operations* 9A-1 (Rocky Mountain Mineral Law Foundation 2005).

10 See *Gerrity Oil & Gas Corp v Magness*, 946 P 2d 913, 926–27 (Colo 1997); *Notch Mountain Corp v Elliott*, 898 P 2d 550, 556 (Colo 1995); *Smith v Linmar Energy Corp*, 790 P 2d 1222, 1224 (Utah Ct App 1990) (mineral estate dominant; mineral owner must exercise rights only as reasonably necessary).

11 *Gerrity*, 946 P 2d 913 at 927 (emphasis added).

12 Colorado Revised Statutes Section 34-60-127.

may allege that the use of the surface was not reasonable or necessary to develop the mineral estate.[13]

In the context of unconventional shale development, the operator generally uses less of the surface area than it would were it engaged in conventional oil and gas development. The use of long horizontal wells allows for the drilling of multiple wells from a single pad. Thus, the overall surface use is more consolidated and less scattered than development of conventional wells. The drilling area for multiple horizontal wells is larger than any single pad used for a conventional well, but the overall affected acreage is typically much less. In addition, the drilling area usually includes production facilities. Again, the size of the facilities will be larger than similar facilities associated with conventional wells, but there are fewer facilities and, thus, less surface impact. Moreover, if gas can be gathered into pipelines to be processed offsite, and water can be delivered to the pads by pipelines, the footprint of a production facility can be reduced even further because tanks and trucks previously needed to store and transport produced minerals and water are no longer part of the equation.

Although unconventional development reduces overall use of the surface area, the hydraulic-fracturing process creates a fairly intensive short-term use of the property. The drilling and hydraulic-fracturing processes require surface space for fracturing equipment and water-handling facilities brought on site for that work. This surface use is temporary, however. Drilling and performing hydraulic fracturing on one long horizontal well typically takes four to 14 days, and that process is multiplied per well; although, on some pads, operators frack multiple wells simultaneously.

Although, as previously noted, surface owners may still bring claims against mineral extractors when shale gas is being developed, some surface owners have recognised the limited impact of unconventional development and have insisted on its use, rather than the use of vertical wells. In *Greeley-Rothe LLC v Anadarko E&P Company LP*,[14] a surface owner planned to develop its surface. The surface owner argued that the accommodation doctrine required the use of horizontal wells because those wells would have less surface impact. The oil and gas operators sought to dismiss the claims, but the court found that the plaintiffs had presented sufficient evidence to survive the motion to dismiss. In particular, the plaintiffs:

> *... provided evidence, construed most favorably to them, that would show that vertical drilling increases truck traffic to and from the wells, requires more roads than would be necessary if the wells were clustered, creates more noise, increases the fire risk, and generally impacts the aesthetics of the land, as well as hampers the ability of the land to be farmed efficiently. These factual showings are sufficient to establish a prima facie case that vertical wells materially interfere with the Plaintiffs' use of the land.*[15]

As discussed in detail below, a surface owner can bring a claim for damages

---

13 See for example, *Coyote Lake Ranch, LLC v City of Lubbock*, No 14-0572, 2016 WL 3176683 (Tex 2016); *Amoco Prod Co v Thunderhead Investments*, 235 F Supp 2d 1163 (D Colo 2002) (federal lands); *EOG Resources Inc v Turner*, 908 So 2d 848 (Miss Ct App 2005).

14 *Greeley-Rothe LLC v Anadarko E&P Company LP*, No 1:08-CV-00401, 2010 WL 1380365 (D Colo 2010).

15 *Id.*

caused to surface lands by oil and gas operations. In addition to these general claims for damages, the surface owner may allege that these damages are a breach of the accommodation doctrine, which, in most of its formulations, requires that the use of the surface to reach the severed mineral estate be 'reasonable' or otherwise minimise damage to the surface. In general, the oil and gas operator must have applied for a permit to drill before the surface owner can seek an injunction, and the surface owner does not have a generic right to specify where oil and gas operations may occur.[16] In *Davis v Devon Energy Production LP*,[17] for example, a surface owner prevented an oil and gas operator from building permanent surfaced roads to facilitate oil and gas operations. The operator sought and won an injunction against the surface owner. The operator argued, and the court agreed, that the construction of permanent roads was reasonable. The roads were necessary to allow access during wet times and when the fields were irrigated, and the road construction was consistent with prior and existing surface roads. Similarly, the Mississippi Court of Appeals found that surface use was reasonable where a road was developed in a manner to minimise the area affected while allowing the use of the road by heavy equipment.[18]

In order to reach the mineral estate, sometimes shale producers desire to drill directionally from a location off-lease, through lands owned by another party, to reach their mineral leasehold. These situations are more frequent in areas where residential or other development limits the availability of surface locations. Disputes can arise regarding which party – the mineral owner, the mineral lessee or the surface owner – has the right to grant permission to pass through the surface. For example, in *Lightning v Anadarko*, the Texas Court of Appeals held that a surface owner has the right to grant an easement through subsurface lands.[19] In this case, Lightning held an oil and gas lease, and Anadarko drilled through the subsurface of the lands covered by Lightning's lease to reach its own lease.[20] Anadarko had permission from the surface owner to drill wellbores through Lightning's leasehold in order to develop Anadarko's adjacent lease.[21] Lightning objected to Anadarko's drilling, but the court found that the surface owner had the exclusive right to allow Anadarko to drill through the subsurface.[22]

It should be noted that there is a basis for a takings claim in the unusual circumstance that a party has a legal right to develop federal minerals underlying a surface estate held by an Indian tribe and such surface use is denied. This type of split estate is common because the United States often put surface lands into trust for tribes but retained the minerals. In *Del Rio v United States*,[23] the US Court of Appeals for the Federal Circuit held that a mineral owner can bring a takings claim against

---

16 *Burkett v Amoco Prod Co*, 85 P 3d 576 (Colo App 2004).
17 *Davis v Devon Energy Production, LP*, 36 S W 3d 419 (Tex App 2004).
18 *EOG Resources Inc v Turner*, 908 So 2d 848 (Miss App 2005).
19 *Lightning Oil Co v Anadarko E&P Onshore LLC*, 480 S W 3d 628, 636 (Tex App 2015).
20 *Id* at 628.
21 *Id.*
22 *Id* at 636.
23 *Del Rio v United States*, 35 Fed Cl 186 (1996) (Del Rio I), withdrawn by 37 Fed Cl 157 (1997) (Del Rio II), withdrawal reversed by 146 F 3d 1358 (Fed Cir 1998) (Del Rio III), reinstated by 46 Fed Cl 683 (2000) (Del Rio IV); 46 Fed Cl 683, 690 (Fed Cl 2000).

the United States in a case in which the United States improperly allows an Indian tribe to preclude the use of tribal surface lands over mineral lands. The court held that such lands are burdened by a surface easement "from birth" and, therefore, the Indian Rights of Way Act does not require tribal consent. As a result, the mineral owner can seek relief for breach of contract against the United States under the Tucker Act. The United States can be held responsible for the tribe's actions where the government gives the tribe a unilateral veto over mineral development, and the mineral owner can bring a claim against the United States for a taking of those minerals.

2.3 Permitting and siting disputes

Disputes regarding governmental approval of drilling activities and the location of drilling activities are common in shale development. On private fee lands, these disputes often concern which government is the proper permitting authority (eg, state or local) and to what extent these governments may regulate drilling activities. On federal land, disputes frequently arise in the context of proper land management and environmental review of proposed mineral leasing. On tribal lands, an additional layer of regulatory oversight – in the form of tribal government – can lead to disputes between operators and the tribes over compliance with tribal rules and permits.

(a) Local versus state regulation of drilling operations

Due to the advent of horizontal drilling and other advanced recovery processes used in shale development, and due to the continued growth of residential development in once-sparsely populated areas, drilling activities increasingly have occurred closer to residential developments. This proximity has led to a variety of disputes over the proper regulatory authority for drilling activities, with both state and local governments (towns, counties, etc) wrestling for power. Disputes often arise when a municipality passes a local law that prohibits oil and gas activities that state law permits (eg, local fracking bans). The primary issues in these disputes are whether the conflicting local law is preempted by state law (because the issue involves one of state concern), and to what extent a local government can exercise its land-use authority to limit or exclude drilling activities within its municipal boundaries.[24]

State and local interests in natural resource development: Many states have comprehensive statutory and regulatory schemes governing oil and gas development

---

24 Although the tension between local government and shale development has occurred primarily in the United States, a recent decision in England shows that this tension is fundamental. The Lancashire County Council refused planning permission to Cuadrilla Bowman Ltd for a series of sites designed to allow exploration for shale development. The Secretary of State for Communities and Local Government reversed that denial (with a minor future condition for one of the sites) in a decision dated 6 October 2016. The secretary of state conducted his review of the council decision under Section 38(6) of the Planning and Compulsory Purchase Act of 2004. The secretary of state considered various concerns about shale development and found that, on balance, shale oil and gas could be developed in a sustainable manner, and that the national policy in favour of shale-gas development carried great weight in this assessment process. Thus, the secretary of state determined that planning permission for the shale exploration process should move forward.

with the goal of promoting efficient, fair and safe development of resources.[25] From a state perspective, uniformity of regulation is necessary to achieve these goals. States are also interested in avoiding 'extraterritorial impacts', which occur when one local jurisdiction takes action that adversely affects residents in another part of the state. In the context of shale development, states try to prevent the actions of one municipality from creating increased production costs, exaggerated production in areas that lack controls, and moratoriums that result in a *de facto* state-wide ban. Municipalities, on the other hand, are more focused on land use within their boundaries and the effects of shale development on local residents. Municipalities have traditionally had authority over land zoning, which can result in the prohibition of industrial activities in certain areas.

Disputes arise when local governments attempt to exercise their zoning authority to limit or ban shale development in a manner that exceeds or conflicts with state law. In these cases, plaintiffs typically assert that state law preempts the local regulation. Other potential claims include takings claims and violations of due process under the federal and applicable state constitutions. As demonstrated in the examples below, courts have typically – but not always – found the local action invalid under the doctrine of preemption.[26]

Local ordinances invalidated: New Mexico: In *Swepi, LP v Mora County, New Mexico*[27] an energy-exploration firm brought a lawsuit against Mora County, New Mexico to prevent the county from enforcing a local law prohibiting corporations from engaging in the extraction of oil, natural gas or other hydrocarbons within Mora County. In challenging Mora County's ban on oil and gas activities, Swepi brought a number of federal constitutional claims, including that the ordinance:

- constituted a taking of private property without compensation;
- violated the federal supremacy clause;
- violated the federal constitutional protections of equal protection and substantive due process;
- violated Swepi's right to free speech; and
- was preempted by New Mexico state law.

The court ruled that Swepi had standing to bring a takings claim under the US Constitution in federal district court because Swepi's oil and gas leases are real property and the local ordinance prohibited Swepi from acting on its leases.[28]

25 See for example, Colorado Revised Statutes Section 34-60-102(a)(1) ("It is declared to be in the public interest to… [f]oster the responsible, balanced development, production, and utilization of the natural resources of oil and gas in the state of Colorado"); North Dakota Century Code Section 38-08-01 ("It is hereby declared to be in the public interest to foster, to encourage, and to promote the development, production, and utilization of natural resources of oil and gas in the state in such a manner as will prevent waste").

26 The doctrine of preemption is the basic constitutional principle that federal law prevails over a conflicting state law covering the same topic and, similarly, that state laws prevail over conflicting municipal laws. For more on this topic, see generally Stephen Elkind, "Preemption and Home-Rule: The Power of Local Governments to Ban or Burden Hydraulic Fracturing", 11 *Texas Journal of Oil Gas & Energy Law* 415 (2016).

27 *Swepi, LP v Mora County, New Mexico,* 81 F Supp 3d 1075, 1151 (DNM 2015).

28 *Id* at 1151.

However, the court ruled that the takings claim was not ripe for review because Swepi had not first sought compensation through an inverse-condemnation proceeding under New Mexico's inverse-condemnation statute.[29] The court also denied Swepi's constitutional equal protection and due process claims.[30] The court did find that parts of the ordinance violated the First Amendment of the federal Constitution by chilling protected speech.[31] Ultimately, the court found that parts of the ordinance impermissibly conflicted with New Mexico state law by banning hydrocarbon-extraction activities.[32] Since these invalid provisions were not severable from the remaining valid provisions of the ordinance, the court struck down the ordinance in its entirety.[33]

Colorado: In the concurrent cases of *City of Longmont v Colorado Oil & Gas Association*[34] and *City of Fort Collins v Colorado Oil & Gas Association*[35] the Colorado Supreme Court considered whether state law preempted a voter-approved fracking ban in the City of Longmont and a moratorium on fracking enacted by the City of Fort Collins. In both cases, the court held that state law preempted the local fracking bans and, thus, invalidated the bans. The court found that the state had a legitimate interest in "the efficient and fair development of oil and natural gas resources" and expressed concern about local governments creating a "patchwork of regulation" that would adversely impact the oil and gas industry.[36] The court further concluded that the local laws cause "serious consequences" felt by those living outside of these towns.[37] The court explained that the ban on fracking inside city limits could increase the cost of fracking in other places in the state.[38] Given the state's legitimate interests, the court ultimately determined that the doctrine of preemption required invalidation of the local fracking bans because those bans conflicted with the Colorado Oil and Gas Conservation Commission's "pervasive oil and gas rules and regulation... [which] include[] a strong interest in the uniform regulation of fracking."[39]

Ohio: In *State ex rel Morrison v Beck Energy Corporation*[40] the City of Munroe Falls sued to prevent Beck Energy from engaging in drilling operations pursuant to its state-issued permit, arguing that Beck Energy's operations violated various local ordinances. Similar to the Colorado cases discussed above, the Ohio Supreme Court found that the state legislature had enacted a "comprehensive regulatory scheme" for oil and gas development.[41] Applying principles of preemption contained in Ohio's

---

29 *Id* at 1158.
30 *Id* at 1146.
31 *Id* at 1187.
32 *Id* at 1200.
33 *Id* at 1088.
34 *City of Longmont v Colorado Oil & Gas Association*, 369 P 3d 573 (Colo 2016).
35 *City of Fort Collins v Colorado Oil & Gas Association*, 369 P 3d 586 (Colo 2016).
36 *City of Longmont*, 369 P 3d at 579.
37 *Id* at 581.
38 *Id.*
39 *Id* at 585.
40 *State ex rel Morrison v Beck Energy Corp* 143 Ohio St 3d 271 (2015).
41 *Id* at 280.

Home Rule Amendment to the state constitution, the court concluded that municipalities may not "discriminate against, unfairly impede, or obstruct oil and gas activities and production operations that the state has permitted."[42] Accordingly, the court held that the local ordinances were invalid.

Pennsylvania: In *Range Resources-Appalachia, LLC v Salem Township*[43] the Supreme Court of Pennsylvania considered whether Salem Township had authority to enact a general ordinance regulating use of the surface and land development associated with oil and gas operations.[44] The court characterised the ordinance as "an attempt by the Township to enact a comprehensive regulatory scheme relative to oil and gas development within the municipality... [that] substantively overlaps with similar regulations" established by the state.[45] Based upon application of preemption, the court invalidated the township's ordinance.

Local zoning ordinances upheld: Despite the weight of authority invalidating local laws based upon state or federal preemption, some courts have upheld local laws that regulate the location of oil and gas exploration and production facilities based on zoning authority. For example, in a case announced the same day as *Salem Township*, the Supreme Court of Pennsylvania upheld use of a zoning district to exclude surface use for natural-gas production.[46] It held that, while a municipality cannot regulate how drilling operations are conducted, it can limit, or even prohibit, the location of operations within the town limits via zoning regulations.[47] Similarly, courts in New York have held that local ordinances may not regulate the "how" of oil and gas operations, but may, in an exercise of zoning, regulate "where" those operations may take place.[48]

(b) Federal permitting disputes

Shale development that occurs on federal public lands is subject to federal leasing permit requirements. The Bureau of Land Management (BLM), an arm of the US Department of Interior, manages public lands under a very complex, multiple-use mandate from Congress, and so must consider multiple factors in land-management decisions, including: industry interests (eg, oil and gas, mining, timber etc); conservation value; and other potential use(s) of the public lands (ie, grazing and recreation).[49] In managing the public lands, the Federal Land Policy Management Act 1976 requires, among other things, that the BLM prevent "unnecessary or undue degradation" of the public lands.[50] The BLM also manages the federal government's

42 *Id.*
43 *Range Resources-Appalachia, LLC v Salem Township*, 964 A 2d 869 (Pa 2009).
44 *Id* at 870.
45 *Id* at 875.
46 *Huntley & Huntley, Inc v Borough Council of Oakmont*, 964 A 2d 855 (Pa 2009).
47 *Id.*
48 *Cooperstown Holstein Corp v Town of Middlefield*, 35 Misc 3d 767 (NY Sup Ct 2012); *Anschutz v Town of Dryden*, 35 Misc 3d 450 (NY Sup Ct 2012). See also *Wallach v Town of Dryden*, 23 NY 3d 728 (NY 2014) (holding that local ban was not preempted by state law).
49 43 USC Section 1701 *et seq.*
50 43 USC Section 1732(b).

mineral interests, including its oil and gas interests. In addition to preventing unnecessary or undue degradation, all major federal actions (including mineral leasing for oil and gas development) are subject to environmental review under the National Environmental Policy Act.[51] Citizen groups and environmental groups frequently challenge decisions to allow oil and gas development on federal lands based upon allegations of inadequate environmental review under the National Environmental Policy Act and failure to prevent unnecessary and undue degradation under the Federal Land Policy Management Act.[52]

Citizen suits against the government: Most federal environmental laws provide opportunities for private individuals and organisations to bring lawsuits under those statutes when the agencies charged with enforcing the laws have not adequately done so, or citizens perceive as much.[53] In the world of shale exploration and production, citizen plaintiffs have focused on governments' and private defendants' failures to abide by federal provisions governing the use of federal land for exploration and production, and failures to follow agency regulations concerning oil and gas facilities. The following are some examples of such suits.

In consolidated cases,[54] plaintiffs Ouachita Watch League and the Ozark Society alleged the US Forest Service and the BLM failed to comply with the National Forest Management Act, the Mineral Leasing Act 1920 and the National Environmental Policy Act because they did not conduct an environmental impact statement and prepare a resource management plan for the leasing of minerals in the Ozark-St Francis National Forest and under Greers Ferry Lake in Arkansas. The Forest Service revised its 1986 Land and Resource Management Plan in 2005 after completing a new environmental impact statement. At that time, the BLM estimated that 10 to 15 new wells would be drilled in the next 10 years. However in 2008, commensurate with the shale-gas boom, the BLM predicted the potential for 1,730 new wells in the forest. To account for this new prediction, the Forest Service prepared a changed conditions analysis and a supplemental information report and concluded that revising the environmental impact statement was unnecessary.[55] Ouachita Watch League claimed the Forest Service failed to take a "hard look" at the changed circumstances and failed to supplement the environmental impact statement.[56] While the Ozark Society argued that, in violation of the National Forest Management Act, the 2005 Forest Service plan did not contemplate the projected increase in oil and gas projects.[57] On defendants' motion to dismiss for failure to state a claim, the US District Court for the Eastern District of Arkansas ruled for the plaintiffs on the

---

51 42 USC Section 4321 *et seq.*

52 43 USC Section 1701 *et seq.*

53 Clean Air Act Section 304, 42 USC Section 7604; Clean Water Act Section 505, 33 USC Section 1365; Resource Conservation and Recovery Act Section 7002; 42 USC Section 6972; the National Environmental Policy Act does not have a specific citizen-suit provision and, thus, claims are generally brought under the Administrative Procedure Act, 5 USC Section 706.

54 See *Ouachita Watch League v Henry* and *The Ozark Society v The United States Forest Service,* 59 F Supp 3d 922 (ED Ark 2014).

55 *Id* at 925.

56 *Id* at 930.

57 *Id* at 933.

National Environmental Policy Act and National Forest Management Act claims, but later granted the defendants' motion for summary judgment on the remaining claims. The case is currently on appeal before the Eighth Circuit.

In *Powder River Basin Resource Council v BLM*,[58] another National Environmental Policy Act case, environmental-group plaintiffs challenged the BLM's review and approval of a resource management plan that included coal-bed natural-gas development and a 16-well drilling-stage project in Wyoming's Powder River Basin.[59] Plaintiffs' complaint for declaratory and injunctive relief alleged that the BLM failed to take the required 'hard look' at direct, indirect and cumulative impacts to the environment before issuing its environmental assessment. Following a systematic review of the BLM's 2011 Draft Buffalo Resource Management Plan and Fortification Creek Management Area Environmental Assessment, the US District Court for the District of Columbia held that the BLM complied with the National Environmental Policy Act and acted reasonably in its methodological review of the potential environmental impacts.

In three consolidated lawsuits brought by the State of New York and environmental and conservation groups against the Army Corps of Engineers,[60] the plaintiffs claimed that the corps should perform an environmental impact study before adopting regulations administered by the Delaware River Basin Commission that would authorise natural-gas development within the Delaware River Basin, a portion of which overlaps the Marcellus Shale formation. Extraction of natural gas there required fracking, and adoption of the final regulations would allow the commission to end an existing moratorium on gas drilling in shale formations under the Delaware River Basin. Defendants filed motions to dismiss for lack of subject-matter jurisdiction, absence of standing and inapplicability of the National Environmental Policy Act. The District Court for the Eastern District of New York ruled in favour of the defendants, holding that the plaintiffs failed to allege an injury-in-fact, and therefore lacked standing; that the action was not prudentially ripe for judicial review; and that the court did not have subject-matter jurisdiction. The court held:

> *While Plaintiffs provide a great deal of support for how their interests may be threatened if natural gas development is allowed in parts of the Basin, as of now it is not allowed, and the mere existence of proposed regulations is not sufficient to allow this court to say Plaintiffs' interests are at risk.*[61]

In apparently the only published opinion in a citizen suit that has focused on hydraulic fracturing,[62] the Center for Biological Diversity argued that the BLM violated the National Environmental Policy Act by insufficiently analysing the

---

58 *Powder River Basin Resource Council v BLM*, 37 F Supp 3d 59 (DDC 2014).

59 The contested land is called the Fortification Creek Planning Area, a 100,655-acre administrative area in northeastern Wyoming. *Id* at 67.

60 *State of New York v US Army Corps*, 896 F Supp 2d 180 (EDNY 2012); The cases are *Delaware Riverkeeper Network, v US Army Corps*, No 11-cv-03780 (EDNY, Aug 4, 2011), *State of New York v US Army Corps*, No 11-cv-03857 (EDNY, Aug 10, 2011) *and State of New York v US Army Corps*, 11-cv-02599 (EDNY, May 31, 2011).

61 *State of New York v US Army Corps*, 896 F Supp 2d at 194.

62 *Center for Biological Diversity v BLM*, 937 F Supp 2d 1140 (ND Cal 2013). There is also a related case, *Center for Biological Diversity v SMR Jewell*, No 13-cv-01749 (ND Cal April 18, 2013).

environmental impacts of fracking when it considered offering 2,700 acres in California's Monterey Shale Formation for four oil and gas leases. The BLM issued a final environmental assessment in which it found "no significant impact" would be created by the lease sale. The environmental assessment briefly addressed fracking and concluded it was not relevant to the analysis of impacts because the issues were outside of the scope of the environmental assessment and not under the BLM's authority or jurisdiction.[63] US Magistrate Judge Paul Grewal of the Northern District of California agreed with the plaintiffs, holding that:

> *Ultimately, BLM argues that the effects of fracking on the parcels at issue are largely unknown. The court agrees. But this is precisely why proper investigation was so crucial in this case. BLM's dismissal of any development scenario involving fracking as "outside of its jurisdiction" simply did not provide the "hard look" at the issue that NEPA requires.*[64]

At time of publication, this case is ongoing: the BLM is preparing a new environmental impact statement and taking other measures related to the leases as required by the Stipulation Regarding Settlement Agreement filed 17 July 2014 with the court.[65]

Finally, in *Citizens for Pennsylvania's Future v Ultra Resources*,[66] the plaintiff alleged that defendant's nitrogen oxide-emitting compressor stations did not have the proper permits under Pennsylvania law and, as such, violated the Clean Air Act. The citizens argued that, although Ultra Resources received G-5 permits, which classified each facility individually, the defendant's facilities should be "aggregated". If aggregated, the eight facilities in question could "potentially emit over 100 [tons per year] of [nitrogen oxide]",[67] whereas they would have been considered emitting less than that amount if the facilities were treated separately under the permit.

Pennsylvania regulations require that the aggregation of multiple air-contamination sources into a "single source" be considered on a case-by-case basis.[68] The court had to determine the meaning of "adjacent" in this case, which would determine if the facilities would be considered a single source. The Pennsylvania court relied upon the Sixth Circuit's opinion in *Summit Petroleum Corp v EPA*[69] to determine the meaning of the term. In *Summit*, the Environmental Protection Agency (EPA) argued that although the petitioner's facilities were physically independent from each other, they were "truly interrelated" under the EPA's Title V program, but the Sixth Circuit disagreed.[70]

After the *Summit* decision, the Pennsylvania Department of Environmental Protection issued guidance for the determination of "adjacent for the purpose of... whether aggregation is appropriate."[71] Under it, the department focuses on the "plain meaning" of the term "adjacent" and, even more specifically, "properties located

---

63 *Id* at 1156.
64 *Id* at 1159.
65 *Center for Biological Diversity v BLM and SMR Jewell*, No 11-cv-06174 (NDCA), ECF No 88-1.
66 *Citizens for Pennsylvania's Future v Ultra Resources*, 11-cv-1360, 2015 WL 769757 (MD Pa 2014).
67 *Id* at *2.
68 25 PA CODE Section 127.201.
69 *Summit Petroleum Corp v EPA*, 690 F 3d 733 (6th Cir 2012).
70 *Citizens for Pennsylvania's Future*, 2015 WL 769757 at *8.
71 *Id* at *9.

within a quarter mile are considered contiguous or adjacent."[72] The guidance also notes that "facilities should not be 'daisy-chained' together to establish a contiguous grouping."[73] The District Court for the Middle District of Pennsylvania held that "the court would be required to 'daisy-chain' a number of properties and their sources in order to find that [Defendant's] facilities constitute a 'major' source."[74] Thus, because the facilities in question were more than a quarter mile apart and operated independently of each other, the court held the facilities did not meet the requirements to be "adjacent" under Summit or the Pennsylvania Department of Environmental Protection guidance.

In recent years, climate change has been a primary concern for environmental groups and will likely continue to be so in years to come. Recently, the Council on Environmental Quality issued its "Final Guidance for Federal Departments and Agencies on Consideration of Greenhouse Gas Emission and the Effect of Climate Change in National Environmental Policy Act Reviews".[75] This guidance presumes that climate change is a real threat to human health and the environment, and describes how federal agencies should consider climate issues when reviewing proposed projects.[76] Federal lawsuits by environmental groups alleging that the government failed to consider properly the effects of climate change in leasing decisions are now making their way through the courts.[77] In the wake of the 2016 US presidential election, however, federal policy regarding natural-resources development and climate change is likely to take a new path towards fewer regulations and more leeway for industry. To the extent that federal policy changes result in increased development of natural resources, it is almost certain that disputes regarding project approvals, environmental standards and the federal government's regulatory responsibilities will play out in the federal courts.

Agency enforcement actions: In addition to citizen-initiated suits, federal agencies can initiate enforcement actions for permit violations or other violations of federal law. Shale operations are subject to a number of federal statutes including: the Comprehensive Environmental Response, Compensation and Liability Act;[78] the

---

72 *Id.*

73 *Id.*

74 *Id* at *11.

75 Christina Goldfuss, Council on Environmental Quality, "Final Guidance for Federal Departments and Agencies on Consideration of Greenhouse Gas Emissions and the Effects of Climate Change in National Environmental Policy Act Reviews", (1 August 2016) available at: https://www.whitehouse.gov/sites/whitehouse.gov/files/documents/nepa_final_ghg_guidance.pdf

76 *Id* at 4 (specifically, agencies should address climate change in their National Environmental Policy Act reviews by: (i) quantifying the project's greenhouse gas emissions based on available data (or to conduct a qualitative analysis when such data is not available); and (ii) considering how changing climate may impact the proposed action (eg, whether the project's location in a flood-prone area will significantly reduce long-term benefits of the project so as to render it worthless)).

77 *Wild Earth Guardians v Jewell*, Case No 1:16-cv-01724 (DDC August 25, 2016). The plaintiff alleged that the BLM failed to weigh the climate impacts for at least 397 oil and gas leases (covering 380,000 acres), and thus violated the National Environmental Policy Act by failing to identify lease sales' direct greenhouse gas emissions, indirect downstream emissions and cumulative emissions when combined with existing onshore development. The plaintiff requested that the court enjoin action on the contested leases until the BLM performs an in-depth analysis of climate impacts from development on public lands.

78 42 USC Section 9601 *et seq.*

Clean Air Act;[79] the Clean Water Act;[80] and the Solid Waste Disposal Act (just to name a few).[81] Federal regulatory and enforcement actions may be brought in federal court or may be heard by the responsible federal agency, subject to later judicial review. The actions may be prompted by an agency's inspection, a complaint or a known violation, or may be brought in the aftermath of an incident or accident. Agencies or third parties may seek monetary relief for clean-up costs, fines or penalties, and the amount sought can vary from a small fine for litter on a well site to a much larger penalty for long-term or widespread environmental contamination. Injunctive relief sought may range from the prohibition of a particular practice by an operator to an attempt to shut down a producer's operations at the state or national level.

EPA actions: In general, the EPA's enforcement initiatives are either civil or criminal. Civil enforcement actions typically result in settlements that include both a monetary penalty and a consent decree or final order. These consent orders are agreed upon by the parties and direct the offending company on how to proceed after the action ends. However, in cases where a defendant knowingly or wilfully committed a serious offence, the EPA often initiates federal lawsuits, typically resolving such matters through a settlement or judgment, and/or fines or imprisonment, as necessary.

As of the end of 2016, the EPA's primary focus has been clamping down on methane emissions from unconventional oil and gas activities. Together with the Department of the Interior, the EPA began taking steps in early 2016 to impose regulations consistent with the Obama administration's initiative to cut methane emissions from the oil and gas industry by 40% to 45% by 2025. The standards and regulations, promulgated pursuant to Section 111 of the Clean Air Act,[82] are the first to target cutting methane emissions from the oil and gas industry. In June 2016, the EPA announced the final rules,[83] which include emission standards and limitations for volatile organic compounds and greenhouse gases for "certain new, modified, and reconstructed equipment, processes, and activities across the oil and natural gas source category."[84] The rules explain the new standards in detail,[85] some of which include monitoring sites for any "fugitive emissions"; if an operator finds any fugitive emissions, that operator must repair that condition within 30 days of discovery.[86] The rules also set standards for equipment to be used in the oil and gas industry; the EPA noted that this will "promote consistency" in enforcement across the industry.[87] Through these new rules, the agency requires owners and operators to confirm all new regulations are met. The rules are clear: "Any owner or operator claiming technical infeasibility, non[-]applicability, or exemption from the

79 42 USC Section 7401 *et seq.*
80 33 USC Section 1251 *et seq.*
81 42 USC Section 6901 *et seq.*
82 See 42 USC Section 7411.
83 81 Fed Reg 35824 – 942.
84 *Id* at 35825 *et seq.*
85 See Id at 35826 – 27.
86 *Id.*
87 81 Fed Reg 35841.

regulation has the burden to demonstrate the claim is reasonable based on the relevant information."[88] In November 2016, the EPA issued an industry-wide information request that formalises aggressive efforts by the agency not only to push industry to collect information, but to identify and repair systematic mechanical and operational issues leading to methane emissions.[89]

The EPA also expressed its current approach to enforcement and regulation of oil and gas companies through its new national enforcement initiatives, announced in mid-February 2016 and which went into effect on 1 October 2016. The new national enforcement initiatives focus on increased transparency in enforcement actions. For example, one of the new provisions in EPA settlement actions now requires entities to share environmental data, including emissions information, with their communities.[90] Another national enforcement initiative adopted by the EPA is "Ensuring Energy Extraction Activities Comply with Environmental Laws".[91] This initiative has the EPA working with states to ensure natural-gas extraction and production is "done in a way that protects the environment and complies with all applicable laws."[92]

Unsurprisingly, the agency focuses on enforcement efforts related to numerous federal environmental statutes' effects on oil and gas development, and it seeks help from federal courts when necessary. For example, in *US v Range Production Company*,[93] Range used horizontal drilling and hydraulic fracturing to drill two gas wells in the Barnett Shale Formation. The EPA claimed that Range's drilling affected two domestic water wells and, after conducting tests on the wells, on 7 December 2010 issued an emergency administrative order under Section 1431 of the Safe Drinking Water Act, violations of which were subject to a statutory civil penalty of up to $16,500 per violation.[94] The emergency order directed Range to remediate impacted areas of the aquifer, provide replacement water supplies and take other action. Range informed the EPA that it disputed the validity of the emergency order and would not comply with some of its terms. On 18 January 2011, the EPA filed a civil enforcement action in the Northern District of Texas seeking injunctive relief and civil penalties against Range. Two days later, Range filed a petition for review of the emergency order in the Fifth Circuit Court of Appeals[95] to determine whether the emergency order was a final agency action and potential violation of Range's due process rights.

The District Court for the Northern District of Texas held that the EPA's order constituted final agency action and was ripe for review, reasoning that: (i) issuing the emergency order to Range and seeking Range's compliance evidenced

88 *Id* at 35844.
89 https://www.epa.gov/controlling-air-pollution-oil-and-natural-gas-industry/oil-and-gas-industry-information-requests (last reviewed 17 December 2016).
90 Andrew Stewart & Jay Rothrock, "EPA Enforcement Initiatives: Agency Puts Oil & Gas In Crosshairs", *American Oil & Gas Reporter*, May Web 2016.
91 "National Enforcement Initiative: Ensuring Energy Extraction Activities Comply with Environmental Laws" (29 February 2016), EPA, available at https://www.epa.gov/enforcement/national-enforcement-initiative-ensuring-energy-extraction-activities-comply.
92 *Id.*
93 793 F Supp 2d 814 (ND Tex 2011).
94 SDWA Section 1431, 42 USC Section 300i, gives the EPA administrator broad authority to act to protect the health of persons in situations where there may be an imminent and substantial endangerment.
95 *Range Resources Corporation v EPA*, No 11-60040. (5th Cir 2011).

"consummation of the agency's decision making process"; and (ii) because the emergency order subjected Range to substantial penalties for non-compliance, and compliance would cause Range to incur significant expense, the order "has the kind of legal and practical effect that is considered to be final agency action".[96] In addition, the court denied Range's motion to dismiss and stayed litigation pending the Fifth Circuit's determination on Range's due process challenge to the order. In the midst of these cases, the Texas Railroad Commission was also conducting tests and holding hearings concerning Range's possible role in the contamination of the water wells. On 22 March 2011, the Texas Railroad Commission issued an order concluding that Range had not caused or contributed to the contamination of the water wells because of its determination that the source of the contamination was from a different shale formation than that tapped by Range's gas wells. The EPA disputed the accuracy of the commission's testing methods and findings. Before the Fifth Circuit had a chance to decide the due-process question, the EPA withdrew the emergency order following the US Supreme Court's decision in *Sackett v EPA*,[97] which held that Clean Water Act administrative orders are final agency actions subject to immediate pre-enforcement review. Thereafter, the parties agreed to dismiss both the Northern District of Texas and Fifth Circuit cases.

State actions: In addition to federal oversight of oil and gas exploration and development, state agencies – many of which are powerful entities designed both to promote oil and gas development and to enforce regulations that implicate environmental, land-use and health concerns related to that development – play a critical local role in enforcing oil and gas regulations against operators.

For example, the Colorado Oil and Gas Conservation Commission has a dual role of promoting Colorado's oil and gas industry while simultaneously regulating it. In January 2016, the commission levied against Benchmark Energy LLC one of the largest fines ever assessed – over $1.2 million – and ultimately revoked its permit. Benchmark Energy LLC had a history of failing to report, investigate or clean up spills.[98] The Colorado Department of Public Health and Environment is also vested with enforcement authority over oil and gas operations in the state. It too recently assessed a significant fine, levying a $2.2 million fine against Suncor Energy for excess benzene emissions from its refinery in Commerce City, Colorado.[99] Similarly, the Texas Railroad Commission regulates the exploration, production and transportation of oil and natural gas in Texas. Its statutory role is to: prevent waste of the state's natural resources; protect the correlative rights of different interest owners; prevent pollution; and provide safety in matters such as those that involve hydrogen sulphide. In September 2015, the Texas Railroad Commission found that Enervest and XTO Energy Inc were not liable for a series of earthquakes from

96 *Id* at 822–823.
97 *Sackett v EPA*, 132 S Ct 1367 (2012).
98 See Colorado Oil and Gas Conservation Commission Order Finding Violation No 1V-564 available at: https://cogcc.state.co.us/orders/orders/1v/564.html (last accessed 19 December 2016).
99 See Colorado Department of Public Health and Environment proceedings *In the Matter of Suncor Energy (USA) Inc*, Compliance Order on Consent, Case No 2011-034 (28 March 2012) and Case No 2011-049 (28 March 2012).

injection activities at their wells in the Barnett Shale.[100] In contrast, the Oklahoma Corporation Commission ordered immediate and indefinite shutdown of 37 wells following a 5.6-magnitude quake in Pawnee County.[101]

(c) *Tribal lands – permitting disputes*

On Indian trust lands, the tribes retain ownership of the mineral resources, including oil and gas.[102] For lands subject to the Allotment Act 1887, the tribes retained the mineral estate and the surface land was subsequently sold to a tribal member or a non-Indian, creating a split-estate. Oil and gas developments on tribal lands are also subject to federal control by the Bureau of Indian Affairs. Complex interactions between tribes, the Bureau of Indian Affairs and industry operators can result in disputes. Tribal governments have begun to regulate oil and gas development through tribal codes, ordinances and constitutions, and, in some cases, through specific, best-management practices. Tribes may also require businesses conducting operations on tribal lands to obtain tribal access permits and business licences. Disputes often arise between companies, tribal agencies and tribal members regarding compliance with the terms with local tribal ordinances and permits.

In addition, disputes often arise when shale development is proposed or occurs on lands that are considered sacred to a particular tribe but are not located within the reservation. For example, in *Solnex v Jewell*, Solnex sued the Department of Interior attempting to obtain the approval for a permit to drill on its lease on lands considered sacred by Blackfeet Nation in Montana.[103] The permit at issue had been pending since 1985.[104] After the judge in the case imposed a deadline to make a decision on the permit application, the Department of Interior cancelled Solnex's lease.[105]

2.4 Commercial disputes

The commercial disputes involved in shale development predominantly involve disputes related to commercial arrangements necessary to produce and sell gas into the market, including gas sales-and-transmission, mineral-leasing and joint operating agreements.

(a) *Gas sales and pipeline transmission*

Typically, gas produced from shale must be processed into its marketable products (eg, methane, ethane, butane and propane). The gas flows from the well as a mixed stream and is transported to a processing plant. Sometimes, the producer sells the gas at the

---

100 Texas Railroad Commission Oil and Gas Docket No 09-0296411 XTO Energy, Inc Proposal for Decision (31 August 2015) and Final Order (3 November 2015) available at www.rrc.state.tx.us/media/31006/09-96411-showcause.pdf (last accessed 19 December 2016). Texas Railroad Commission Oil and Gas Docket No 09-0296410 Enervest Operating, LLC, Proposal for Decision (10 September 2015) and Final Order (3 November 2015) available at: www.rrc.state.tx.us/media/31005/09-96410-showcause.pdf (last accessed 19 December 2016).

101 See www.occeweb.com/News/2016/11-23-16EARTHQUAKE%20ACTION%20SUMMARY.pdf (last accessed 19 December 2016).

102 See *United States v Shoshone Tribe*, 304 US 111 (1938) (holding that when lands are reserved or otherwise set aside for the tribes in executive orders, treaties or agreements approved by Congress, the tribes hold the beneficial rights to the soil and the mineral interests under the lands).

103 See *Solnex LLC, v Jewell*, No 1:13-cv-993 (DDC 2013).

104 *Id.*

105 *Id.*

wellhead to the gas processor, who then resells the products into the market. At other times, the producers themselves sell the gas into the market after processing. The terms of gas-processing and purchase-and-sale agreements vary because each contract addresses the particular needs of a party, as well as each party's bargaining position, the relevant market and other related factors. However, as discussed below, there are some common provisions that frequently lead to disputes between gas producers and midstream companies (the companies that process, transport and sell the gas into the market). In some cases, the gas purchase agreements include dispute-resolution clauses that require a party to provide the other party notice of the dispute and also require the parties to attempt to reach resolution before filing a formal claim.

Supply commitments and take-or-pay obligations: Since gas gathering, processing and transportation require significant capital investment in the related infrastructure, many gas-processing and -sales agreements have minimum supply obligations. If the producer is unable to fulfil those commitments, it may be subject to a penalty. Similarly, a contract may obligate a gas purchaser to purchase a minimum amount of product. If the purchaser is unable to take the product, it must still make payment (ie, take-or-pay contracts). When market forces or other events make delivery of product minimums or acceptance of minimum product amounts difficult, parties sometimes attempt to invoke the concept of *force majeure* to relieve their minimum obligations under the contract. Whether or not the particular condition is an event of *force majeure* or whether the related doctrines of commercial impracticability/impossibility excuse a party's obligations is often litigated.[106]

Proper payment and audits: Most gas-processing and -sale agreements contain provisions allowing the parties to audit the books and records of the other party over certain time periods (typically looking back two or three years prior to the audit request). From the gas producer's perspective, these provisions are important to ensure that it is being properly paid for gas deliveries. Upon completing an audit of the deliveries and sales of gas, a producer may believe that it has been underpaid. Disputes arise regarding a number of issues, including: (i) whether the gas delivered was properly allocated to the producer's account; (ii) whether the processor maintained the proper pressure in the gathering lines so that the producer could make its deliveries; and (iii) whether the processor rightfully retained part of the gas stream pursuant to the contract.[107]

---

106 See *Great Lakes Gas Transmission Limited Partnership v Essar Steel Minnesota LLC*, 871 F Supp 2d 843 (D Minn 2012) (holding that the doctrine of commercial impracticability/impossibility did not relieve buyer of its contractual duty to pay for transportation of gas), vacated and remanded on other grounds, No 16-1101, 2016 WL 7046743 (8th Cir Dec 5, 2016); *Virginia Power Energy Marketing, Inc v Apache Corp*, 297 SW 3d 397 (Tex App Houston [14th Dist] 2009, pet denied) (holding that Apache appropriately declared *force majeure* when Hurricane Katrina/Rita disrupted its ability to deliver gas at the two delivery points named in the governing contract).

107 See for example, *PDC Energy v DCP Midstream*, Case No 14-cv-01033 (D Colo 2014) (plaintiff alleged various underpayments after audit of DCP Midstream); see *Dynegy Midstream Sers, Ltd P'ship v Apache Corp*, 294 SW 3d 164, 168-69 (Tex 2009) (plaintiff sought recovery for "unaccounted-for" gas that disappeared between production at Apache's wellheads where title to all gas transferred to the processor and sale to customers at the processing facility).

Price review provisions: Some gas contracts contain provisions that allow the parties to renegotiate the sales price or processing/transportation service price under certain conditions (price reopeners). Price reopeners are typically triggered upon specified events or changes in circumstance. Disputes can arise over whether the reopening provision was properly triggered and what the ultimate price should be.[108]

(b) Royalty holder disputes

Owners of a mineral interest typically lease that interest to shale producers for development. In exchange, mineral owners receive a royalty on produced minerals that are sold in the market. These relationships are governed by the parties' particular oil and gas lease. These leases typically require production in commercial quantities within a certain number of years (the primary term) and continue for as long as such production continues. Courts have also held that oil and gas leases contain various implied duties unless they are expressly disclaimed. One such duty – the 'duty to market' – has been the frequent subject of litigation between royalty holders and shale producers.

The shale producer generally has a duty to market the gas once it has been discovered, including the duty to obtain the best price. There has been extensive litigation on the subject of whether the duty to market requires the shale producer to bear alone the costs of putting the gas into marketable condition (ie, gas processing costs, as discussed above). Courts in some jurisdictions have interpreted the duty to market to require the producer to bear the cost of putting the gas into marketable condition for sale. These jurisdictions include Colorado, West Virginia, Oklahoma, and Kansas.[109] Other states, such as Texas and North Dakota, allow producers to deduct downstream costs from the proceeds on which they pay a royalty.[110] While these states have default rules regarding how post-production costs are allocated in leases that are silent on this issue, it is important to remember that the parties may themselves allocate these costs in their particular leases.[111]

Allocation of post-production costs has been perhaps the most extensively litigated issue between royalty holders and producers, with some of these cases reaching the state supreme courts. Other disputes that have received less attention include conflicts over whether leases terminated based on the production activities of the lessee, and whether royalties have been accurately calculated and timely paid.[112]

---

108 See for example, *Gas Natural Aprovisionamientos, SDG, SA v Atlantic LNG Company of Trinidad and Tobago*, 2008 WL 4344525, at *1 (SDNY 2008) (upholding a price formula imposed by an arbitration panel based upon the price reopener clause in the governing sale and purchase agreement).

109 See John Burritt McArthur, "Mineral Royalties, Deductions, and Fawcett v Opik: Continuity and Change in the Revised-but-Still-Standing Kansas Marketable-Product Rule", 64 *University of Kansas Law Review* 63, 65 (2015) ("[t]he supreme courts in Colorado, Oklahoma, and West Virginia have joined the Kansas Supreme Court in authoring the leading marketable-product decisions").

110 *Heritage v NationsBank*, 939 SW 2d 118 (Tex 1996), reh'g denied, 960 SW2d 619 (Tex 1997); see *Bice v Petro-Hunt, LLC*, 768 NW 2d 496, 502 (ND 2009) ("We join the majority of states adopting the 'at the well' rule and rejecting the first marketable product doctrine").

111 *Chesapeake Expl, LLC v Hyder*, 483 SW 3d 870, (Tex 2016). In *Hyder*, the court held that where a lease called for "'a perpetual, cost-free (except only its portion of production taxes) overriding royalty of five percent (5.0%) of gross production obtained' from direction wells drilled on the lease but bottomed on nearby land", that the parties thereto had agreed that post-production costs should not be deducted from the amount on which the royalty was paid.

2.5 Joint operating agreements

Joint operating agreements govern the relationship between working interest owners or mineral owners whose interest contributes to a unit. The joint operating agreement provides the contractual basis for the cooperative exploration, development and production of oil and gas properties among multiple parties. The underlying purpose of the joint operating agreement is to provide a structure for contributing capital, allocating risk and designating operator responsibility. Under a typical joint operating agreement, the operator will assume the operating expenses and then bill the non-operators for their proportionate share. As in any commercial relationship, disputes may arise and are influenced by market conditions or other factors. These disputes may concern, among other issues: the propriety of the operator's conduct; accounting of expenditures; gas balancing; issues as to the working-interest owner's proportionate share of production; and the ability of working-interest owners to exercise preferential purchase rights contained in most form joint operating agreements.

2.6 Tort cases against oil and gas companies

Although land-use, contract and regulatory issues permeate the many disputes surrounding the extraction of petroleum from shale, personal-injury allegations have also occupied substantial court time. Hydraulic fracturing (and its often-attendant misconceptions of the facts and the law) has borne the brunt of plaintiffs' claims.

For example, in *Cerny v Marathon Oil Corp*,[113] homeowners who had leased the mineral rights on their land to Marathon filed suit in Texas alleging that Marathon's fracking activities in the Eagle Ford Shale damaged their health and their property. They brought claims for private nuisance, negligence, gross negligence and negligence per se. The Texas state court granted summary judgment for defendants, and plaintiffs appealed. In their appeal, the plaintiffs disclaimed any personal-injury damages but asserted that they were seeking damages for medical expenses, home repair and loss of home value due to sinkholes, pollution, noxious odours and dead plants and animals on their property. The plaintiffs appealed two main issues: (i) whether the trial court erred in granting the defendants' motions for summary judgment; and (ii) whether the trial court "abused its discretion in striking the majority of the plaintiffs' summary judgment evidence" as "inadmissible hearsay, unqualified lay opinions, and unreliable, speculative, and conclusory expert opinions."[114] Although the plaintiffs argued that because they "disclaimed recovery for any 'personal injury damages'"[115] their claims did not "invoke *Merrell Dow Pharms. v Havner*",[116] the appellate court applied the holding of *Havner*[117] to determine the strict causation standards as applied to plaintiffs' claims. These standards require

112 See for example, *Rook v James E Russell Petroleum, Inc* 679 P 2d 158 (Kan 1984) (discussing whether lease terminated due to abandonment); *Parry v Amoco Production Co*, 2003 WL 23306663 (2003) (D Colo 2003) (discussing whether lessee improperly included certain deductions when calculating royalties).

113 *Cerny v Marathon Oil Corp*, 480 SW 3d 612 (Tex App 2015).

114 *Id* at 617.

115 *Id* at 618.

116 *Id*.

117 *Merrell Dow Pharms v Havner*, 953 SW 2d 706 (Tex 1997).

expert "testimony... to demonstrate that exposure 'more likely than not' caused the injury by pointing to at least two epidemiological studies demonstrating a statistically significant doubling of the risk as proof of general causation."[118] The court held that the plaintiffs "suffered from multiple chronic health conditions that existed prior to the defendants' commencement of oilfield operations" and that "the Cerny's home had foundation damage prior to the defendants' operations."[119]

Havner also requires plaintiffs to promulgate evidence to exclude any other potential causes of plaintiffs' alleged damages and injuries. According to the appeals court, plaintiffs' experts failed to do so, and "other companies' emissions [were] plausible potential causes of the [Plaintiffs'] personal injuries and property damage."[120] Since there were multiple companies working near the plaintiffs' property, the court held that plaintiffs must have proven that the defendants caused the alleged damages, and they failed to do so. Therefore, the appeals court affirmed. Plaintiffs appealed to the Texas Supreme Court and the court denied review on 2 December 2016.[121]

In one of the first hydraulic-fracturing cases in the country, which eventually settled after remand, *Antero Resources Corporation v Strudley*,[122] a family of four alleged that a natural-gas exploration-and-production company and its vendors, whose operations occurred near their home, contaminated the family's air and water, thereby causing health and property damage. As the plaintiffs' case appeared to be fundamentally factually anaemic, the defendants' counsel deployed a progressive technique to leverage dismissal, called a 'Lone Pine' motion. Lone Pine motions stem from a New Jersey Superior Court case in which homeowners sued operators of a landfill in a multiple-party case for property damage and personal injury.[123] In that case, the court determined that the defendants would be unfairly burdened by the tasks and expenses of discovery unless the plaintiffs could demonstrate as an initial matter that their case was well substantiated by fact. Lone Pine orders are usually issued after initial disclosures but before discovery, and require plaintiffs in toxic-tort cases to provide evidence sufficient to establish a *prima facie* case of injury, exposure and causation, or else face dismissal of their claims. Federal Rules of Civil Procedure 16(c) authorises the use of Lone Pine orders in complex federal cases, but they had not been used in Colorado before *Strudley*. The defendants asked the trial court to require the family to present *prima facie* evidence that they suffered injuries attributable to the drilling operations, and the court issued a Lone Pine order forcing the plaintiffs to do so. The plaintiffs failed to make the required showing pursuant to the order, and the trial court dismissed the case. The Colorado Court of Appeals reversed, holding that Lone Pine orders "are not permitted as a matter of Colorado law".[124] In a myopic blow to the defence bar, the state's highest court agreed:

---

118 *Cerny v Marathon Oil Corp*, 480 SW 3d at 621.
119 *Id.*
120 *Id.*
121 *Cerny v Marathon Oil Corp*, No 16-0011 (Tex 2016).
122 *Antero Res Corp v Strudley*, 347 P 3d 147 (Colo 2015). Note that Andrew Lillie, co-author of this chapter, Daniel Dunn, his partner at Hogan Lovells US LLP, and others represented Antero Resources in this matter.
123 *Lore v Lone Pine Corp*, 1986 WL 637507 (NJ Super Ct 1986).
124 *Strudley v Antero Res Corp*, 350 P 3d 874, 877 (Colo App 2013).

> *Although the comments to Colo. R. Civ. P. 16 promote active judicial case management, the rule does not provide a trial court with authority to fashion its own summary judgment-like filter and dismiss claims during the early stages of litigation. Colorado law already provides sufficient protection against frivolous or unsupported claims and burdensome discovery under existing rules and procedural safeguards.*[125]

One of the first jury verdicts in which exploration and production activities were implicated as causing plaintiffs' injuries was handed down in the Dallas County Court in Texas in 2015, in *Parr v Aruba Petroleum, Inc.*[126] The plaintiffs' home was located near defendants' operations. The plaintiffs brought claims for negligence, negligence *per se*, gross negligence, private nuisance, strict liability and trespass. The plaintiffs claimed that numerous gas operations near their home caused serious health problems for the family. At trial, only one claim was presented to the jury: private nuisance. The jury found for the plaintiffs and awarded $275,000 for the plaintiffs' loss of market property value, $2 million for past physical pain and suffering, $250,000 for future physical pain and suffering, and $400,000 for past mental anguish. The jury found that the energy company's activities were "intentional", thereby creating a "private nuisance" under Texas law.[127] Defendant appealed and, as of publication of this chapter, the case is pending.[128]

Numerous lawsuits regarding shale oil and gas extraction operations have involved the question of whether the activities are "abnormally dangerous", thereby giving rise to strict liability. For example, in *Ely v Cabot Oil & Gas Corp,*[129] the plaintiffs, a group of property owners that had entered into separate leases with Cabot Oil & Gas Corporation for the purpose of exploring and drilling for oil and gas, sought damages as a result of drilling and hydraulic-fracturing activities that allegedly occurred within 1,000 feet of their drinking-water wells. Plaintiffs claimed Cabot's unconventional drilling activities "constitute[d] abnormally dangerous activities under Pennsylvania law, and therefore should be subject to strict liability."[130] The court disagreed, holding that "for decades, courts have uniformly refused to find that oil and natural gas drilling and related activities are ultra[-]hazardous or abnormally dangerous",[131] and therefore, they are not subject to strict liability.[132] The court explained that the plaintiffs failed to submit evidence showing that even with proper due care the risks associated with unconventional well drilling and completion cannot be eliminated. Indeed, the court found the record showed that "risks are substantially mitigated when due care is exercised."[133] The court adopted the defendants' position that "risks from a properly drilled, cased and hydraulically fractured gas well are minimum."[134] The court also noted, unsurprisingly, that, like any industrial activity, drilling and fracking activities may

125 *Antero Res Corp v Strudley* 347 P 3d at 151.
126 *Parr v Aruba Petroleum, Inc,* No CC-11-01650-E (Tex C'nty Ct Apr 22, 2014).
127 *Id* at Jury Verdict.
128 *Aruba Petroleum, Inc v Parr,* No 05-14-01285-CV, 2015 WL 7308038 (Tex App Dallas Nov 9, 2015).
129 *Ely v Cabot Oil & Gas Corp,* 38 F Supp 3d 518 (MD Pa 2014).
130 *Id* at 520.
131 *Id.*
132 *Id* at 530.
133 *Id* at 531.
134 *Id* at 529.

present risks if not properly undertaken, but with proper care those risks are outweighed by the economic benefits of oil and gas production.

In *Berish v Southwestern Energy Production Co*,[135] the plaintiffs, a group of residents living near wells owned and operated by the defendants, Southwestern Energy Production Co and Southwestern Energy Co, brought claims for violation of the Pennsylvania Hazardous Sites Cleanup Act,[136] negligence, private nuisance, strict liability, trespass and the need for the establishment of a medical-monitoring trust fund. The plaintiffs alleged that the defendants' natural-gas well, located in Susquehanna County, Pennsylvania, was insufficiently cased, which allowed fracking solution to contaminate the plaintiffs' drinking-water supplies. This allegedly caused health problems and lowered the value of the plaintiffs' properties. The defendants filed a motion to dismiss the claims for strict liability and emotional distress.

The US District Court of the Middle District of Pennsylvania applied Sections 519 and 520 of the Restatement (Second) of Torts to decide the strict-liability claim, and held that although the plaintiffs may have problems meeting certain prongs of Section 520, there is no requirement under the Federal Rule of Civil Procedure that the plaintiffs must have demonstrated all of the facts underlying the claim before discovery occurred. The court found that although hydraulic fracturing had never been considered an "abnormally dangerous activity" to which strict liability would be applied, the determination was "fact-intensive" and required the court to "wait until discovery is complete before making this determination".[137] Therefore, the court denied the motion to dismiss the strict-liability claims.

With regard to the emotional-distress claim, the court noted that under Pennsylvania law, a plaintiff must demonstrate physical injury to claim damages for emotional distress. Since only one plaintiff pleaded facts amounting to physical injury, all claims for damages brought by the other plaintiffs were dismissed. However, the court granted leave to the plaintiffs to file an amended complaint to claim damages for inconvenience and discomfort. The case was eventually settled and dismissed.

In 1988, Pennsylvania passed a Hazardous Sites Cleanup Act, which provides the Pennsylvania Department of Environmental Protection the funds and authority for the cleanup of hazardous waste sites, the authority to investigate liability and effect cost recovery, and the authority to enforce the Hazardous Sites Cleanup Act. Many cases in the Middle District of Pennsylvania court have alleged violations of the Hazardous Sites Cleanup Act. For example, in *Fiorentino v Cabot Oil & Gas Corp*,[138] property owners, who executed leases with defendants, alleged natural-gas drilling and fracking released methane, natural gas and other substances into their water and onto their land. The defendants, Cabot Oil & Gas Corporation and Gas Search Drilling Services Corporation, filed a motion to dismiss the plaintiffs' claims that the

---

135 *Berish v Southwestern Energy Production Co*, 763 F Supp 2d 702 (MD Pa 2011).
136 35 PS Sections 6020.101–6020.1305
137 *Id; Berish*, 763 F Supp 2d at 705.
138 *Fiorentino v Cabot Oil & Gas Corp*, 750 F Supp 2d 506 (MD Pa, 2010); see also *Ely v Cabot Oil & Gas Corp*, 38 F Supp 3d 518 (MD Pa 2014).

defendants violated the Hazardous Sites Cleanup Act, and for strict liability, establishment of a medical-monitoring fund, gross negligence and negligence *per se*.

The plaintiffs sought relief under Section 702 of the Hazardous Sites Cleanup Act, which sets forth the costs for which the individual responsible for a release of a hazardous substance "is strictly liable".[139] The court agreed that the plaintiffs stated plausible claims under Section 702 and denied the defendants' motion to dismiss that claim. With regard to the strict-liability claim, the plaintiffs pointed out that Pennsylvania courts have only concluded that "storage and transmission of gas and petroleum products" are not abnormally dangerous but have not concluded whether drilling and other operations are abnormally dangerous.[140] The court agreed and did not grant defendants' motion to dismiss that count because it would be "improvident to automatically extend this reasoning to drilling activities without more thorough consideration".[141]

With regard to the medical-monitoring claim, the court held that the plaintiffs sufficiently alleged plausible facts to support the claim that defendants' operations, which involved the use of fracking fluids that contain "hazardous chemicals that are carcinogenic and toxic", caused health issues consistent with constant toxic exposure.[142] In addition, the court denied the defendants' motion to strike certain damages allegations because plaintiffs adequately alleged actual physical injury in their claims for fear of future illness and emotional distress. Finally, the court denied the defendants' request to strike punitive damages allegations because the plaintiffs sufficiently alleged that "[d]efendants were grossly negligent in the operation and drilling[,] thus subjecting Plaintiffs to many harms."[143] All but three of the plaintiffs and the defendants eventually settled. The three remaining plaintiffs continued their suit as *Ely v Cabot Oil & Gas Corp*.[144]

The plaintiffs in *Roth v Cabot Oil & Gas Corp*[145] brought claims for negligence, negligence per se, private nuisance, strict liability, trespass, inconvenience and discomfort, breach of contract, fraudulent misrepresentation and inducement, and violation of the Hazardous Sites Cleanup Act related to drilling and hydraulic fracturing that occurred within 1,000 feet of the plaintiffs' property. Cabot and its co-defendants moved to dismiss.

The plaintiffs entered into a lease with the defendants for the right to drill and extract natural gas. Part of the lease specified that the defendants would test the plaintiffs' water supply before and after drilling and exploration activities to ensure it was not adversely affected by those activities. The defendants had also agreed to share the full results of these tests with the plaintiffs, and to immediately inform the plaintiffs if the defendants' activities disrupted or adversely affected the plaintiffs' property, and to take the necessary steps to return the property to "pre-drilling conditions".

Four months after operations began, the plaintiffs noticed their water contained

139 35 PS Section 6020.702.
140 *Fiorentino*, 750 F Supp 2d at 512.
141 *Id*.
142 *Id* at 513.
143 *Id* at 515.
144 38 F Supp 3d 518 (MD Pa 2014).
145 *Roth v Cabot Oil & Gas Corp*, 919 F Supp 2d 476 (MD Pa 2013).

sediment, was brown and emitted a foul odour. An inspection of one of the defendant's well pads showed that the waste pit liner was full of holes, which permitted hazardous waste water to enter the soil, leading to contamination of the groundwater. Pennsylvania Department of Environmental Protection representatives cited the defendants for negligent cement and casing work, and the spillage of half a barrel of waste fluid on the surface of a well pad. Eight months after the drilling activities began, the Pennsylvania Department of Environmental Protection tested the plaintiffs' water and deemed the water unfit for human consumption.

In rendering its decision on the defendants' motion to dismiss, the District Court for the Middle District of Pennsylvania held that the inferences drawn from the record were "sufficient to establish that the defendants plausibly released a 'hazardous substance' into the plaintiffs' groundwater supply."[146] The court also held that the timing of when drilling activities began and when the plaintiffs noticed a change in their groundwater was sufficient to establish that defendants caused the contamination. Regarding the negligence claim, the court found that the plaintiffs' "temporal and physical proximity" to the defendants' activities, as well as the lack of alternative causes, permitted "the reasonable inference that the Defendants were responsible for that harm."[147]

For negligence per se, the court followed the precedent set forth by the Western District of Pennsylvania in *Fallowfield Development Corp v Strunk*,[148] in determining whether the Hazardous Sites Cleanup Act, the Pennsylvania Solid Waste Management Act or the Pennsylvania Clean Streams Law can serve as bases for negligence *per se*. In *Fallowfield*, the court held that plaintiffs were "incidental but unintentional beneficiaries"[149] of the law, and thus could not use the Clean Streams Law as claims for negligence *per se*. However, the court agreed that the intent behind the Solid Waste Management Act and the Hazardous Sites Cleanup Act is to "protect the citizens of this Commonwealth from the dangers of the improper disposal of hazardous and solid waste… [and that] these policies would be furthered by allowing violations of the [Solid Waste Management Act] and [the Hazardous Sites Cleanup Act] to serve as the basis for a claim of negligence *per se*."[150] The court also denied the motion to dismiss the count of private nuisance. It held that the plaintiffs sufficiently alleged that the defendants "caused a substantial invasion to the Plaintiffs' interest in the private use of their Property causing a 'seriously annoying or intolerable' nuisance, which satisfied the second prong of Section 822 of the Restatement (Second) of Torts."[151] For strict liability, the court deferred its decision to the summary-judgment stage, after a factual record would be better developed. The court dismissed the trespass, inconvenience and discomfort, and fraudulent misrepresentation claims. In November 2013, the parties settled their dispute and the case was dismissed with prejudice.

146 *Id* at 485.
147 *Id* at 487.
148 1990 WL 52745 (WD Pa Apr 23, 1990).
149 *Id* at 488 (quoting *Fallowfield*).
150 *Id* at 488–89 (quoting *Fallowfield*).
151 *Id* at 491.

Trespass claims have also played a starring role in shale oil and gas tort cases. For example, in *Coastal Oil & Gas Corp v Garza Energy Trust*,[152] the plaintiffs, a group of royalty-interest owners of a natural-gas lease, brought claims against operators for subsurface trespass, breach of the duty of good faith and fair dealing, and breach of implied covenants, including the duty to protect against drainage. The Texas Supreme Court held that gas drainage between properties caused by hydraulic fracturing is not a form of trespass, but is sanctioned by the "rule of capture", holding that "the rule of capture determines title to gas that drains from property owned by one person onto property owned by another."[153] The court explained that application of the rule of capture was "justified" in that case because an individual can protect himself from drainage by drilling his own well, which can prevent the migration of gas.[154] However, the court clarified that the rule of capture cannot be used "to shield misconduct that is illegal, malicious, reckless, or intended to harm another without commercial justification."[155]

2.7 Conclusion

Whether the disputes rest in contract or tort, involve a permit violation or a land-use conflict, or concern a single family, a group of companies or an entire state, arguments over unconventional oil and gas exploration, drilling, fracking, production and sale present myriad legal challenges and obstacles. As the shale revolution spreads to other parts of the world, expect the lawyers to follow.

3. Future litigation in Europe

The litigation in the United States previously described in this chapter reflects North America's position at the forefront of global shale gas development. Similar types of dispute are likely to follow as the industry develops in other parts of the world. After North America, Europe currently presents the most likely geographical location for such disputes to arise.

Although the underlying subject matter of shale gas disputes in Europe is likely to be the same as in the United States, the causes of action and potential remedies will differ, as will the regulatory environment. One major difference between the United States and Europe is that Europe is unlikely to experience similar contractual disputes between developers and landowners relating to the ownership of estates and mining rights. In Europe, the sub-soil mineral rights are usually owned and licensed by the State, which imposes terms on the development of the resources, including as to rates of royalty payable to the State. Therefore, the relationship between a developer and a landowner on whose land operations are taking place will generally be non-contractual in nature.

The main non-contractual causes of action (and likely remedies) in England and Poland are briefly outlined below. These are two countries in Europe which, at the time of writing, have seen shale gas exploration take place. Although the

---

152 *Coastal Oil & Gas Corp v Garza Energy Trust*, 268 SW 3d 1 (Tex 2008).
153 *Id* at 14.
154 *Id*.
155 *Id* at 17.

development of the shale gas industry in both countries remains embryonic (and questions have arisen in recent years as to the commercial and geological viability of Poland's shale plays), the legal systems of these countries illustrate the main differences between the common law and civil law traditions that exist in Europe and are therefore considered here. Whereas England has a common law, case-based tradition (very similar to that in the United States), Poland is a civil law system with codified laws, which is similar to that in most continental European countries.

3.1 Likely causes of action

(a) England

The tort litigation arising out of or in relation to fracking in England is likely to be similar to that in the United States. As in the United States, the fracking process and related industrial activity, damage caused to land as a result of a nuisance, the escape of fracking fluid, water contamination, noise or air pollution are likely to result in actions in nuisance and/or trespass. Similarly, harm to human health is likely to give rise to personal injury actions in negligence.

As in the United States, there are causes of action for private and public nuisance, although due to the higher thresholds required to found an action in public nuisance, actions in private nuisance are likely to be more common.

Private nuisance arises where the conduct of a person: (i) causes an encroachment on a neighbour's land; (ii) causes physical damage to a neighbour's land, buildings, works or vegetation on it; or (iii) unduly interferes with a neighbour's comfortable and convenient enjoyment of his land.

Liability for damages caused by nuisance will arise only if the damage was reasonably foreseeable. Generally, a complained-of nuisance must be a real interference with the comfort or convenience of living judged by the standards of an average person and taking account of all relevant circumstances. However, where the nuisance causes material damage to property, or reduces its value, the surrounding circumstances do not have to be taken into account and the nuisance will be actionable. The normal measure of compensatory damages in nuisance is the value of the loss resulting to the party as a natural consequence of the wrongful act of the person causing the nuisance. Where damage to property is involved, the measure of damages is the difference in monetary value to the claimant's interest in the property before and after the damage was done, not the cost of repair (although the repair costs can be a relevant factor in assessing the loss of value). In addition to damages, an injunction which orders the offending activity to cease is a common remedy for nuisance.

Statutory nuisance actions can also be commenced under the provisions of the Environmental Protection Act 1990 by governmental authorities. The consequences of a statutory nuisance are likely to be abatement notices against the nuisance creator, followed by a fine for non-compliance.

An important recent court decision has clarified the English law of nuisance in circumstances where a potentially nuisance-causing activity is carried out pursuant to the terms of a statutory licence. In *Barr v Biffa Waste Service Ltd*, the defendant operated a waste site pursuant to the terms of a permit issued by the Environment Agency. The terms of licence included a requirement that steps be taken to control and minimise odours from the activities. Residents living close to the waste site brought an action in nuisance complaining about odours from the site. At the first instance, the court decided, in effect, that nuisance would not arise provided that the operator of the site operated it in accordance with the terms of its licence. This decision was overturned by the court of appeal, which decided that nuisance could arise notwithstanding compliance by the operator with the licence. This decision is likely to have a direct impact on shale gas developers who will be carrying out shale gas development activities under the terms of a hydrocarbon on-shore development licence.

As in the United States, actions for personal injury relating to harm to humans, or for damage to property not involving nuisance, are founded in negligence. Negligence will arise where a defendant breaches a 'duty of care' owed to the claimant which causes reasonably foreseeable damage. Compensatory damages seek to put the claimant in the same position as if the breach of duty had not occurred. Also, where the alleged breach is continuing, it may be possible to obtain an injunction to put a stop to the offending activity.

In an early reflection of the permitting and siting disputes experienced in the United States, England has already experienced disputes connected with the granting of planning permission by relevant local authorities for shale gas drilling activities. As noted previously,[156] the decision to deny planning permission to Cuadrilla Bowman Ltd by Lancashire County Council was overturned by the Secretary of State for Communities and Local Government. That planning decision has now been appealed by a local community action group pursuant to Section 288 of the Town and Country Planning Act 1990, on the basis that the decision was taken unlawfully by reason of a failure to correctly understand and apply the development plan and to act in accordance with the rules of natural justice. At the time of writing, the decision is awaited.

In a separate attempt to overturn planning permission granted by a local authority, the environmental group Friends of the Earth and a local action group sought a judicial review by the English High Court of North Yorkshire County Council's decision to grant permission to Third Energy to drill for and produce shale gas using hydraulic fracturing in its concession area in Ryedale. The application was essentially made on the basis that the local council had failed to properly take into account the impact of the production and use of any gas from the well at the local Knapton power station on greenhouse gas emissions when assessing the grant of planning permission. The application was rejected in December 2016.[157]

(b) *Poland*

As previously mentioned, Poland is a civil law system with a codified system of laws.

156 See footnote 25.

157 *(1) Friends of the Earth (2) Frack Free Ryedale v North Yorkshire County Council and others* [2016] EWHC 3303 (Admin).

The main acts which give rise to relevant causes for action are: (i) the Geological and Mining Law dated 9 June 2011; and (ii) the Civil Code of 23 April 1964 (which shares many similarities and is influenced by the civil codes of Germany, France and Switzerland).

Pursuant to the above statutes, mining damage arises where: (i) an event occurs during the operation of a mining plant (which includes gas production); (ii) damage occurs; and (iii) there is a causal link between the event and the damage.

The Geological and Mining Law stipulates that a property owner may not object to the threat caused by the operation of a mining plant if such operation is carried out in compliance with the Geological and Mining Law. However, such an owner may demand that the damage caused by the operation be redressed.

Redress for any damage is made by restoration of the property to its previous condition. Damage to agricultural and forest land should be redressed in the manner specified in the provisions relating to the protection of such land. Alternatively, the party that has suffered damage may, upon consent of the wrongdoer, carry out the necessary remedial activity in return for payment of an appropriate sum of money.

If the operation of a mining plant is not carried out in compliance with the Geological and Mining Law, the injured party may also rely on the provisions of the Civil Code described below.

The common law concept of nuisance has its equivalent in Article 144 of the Civil Code, which provides that a property owner shall abstain from acts that excessively disturb the use of neighbouring properties. In order to prove a breach of Article 144, it is necessary for a claimant to demonstrate that the operation of a mining plant is not carried out in compliance with the Geological and Mining Law and affects the use of his property beyond a normal scope, arising from the social and economic purpose of the real property and local relations. The property owner is entitled to request that use of the land should revert to its lawful status and for breaches of the law to be stopped.

Chapter 6 of the Civil Code addresses torts, including harm to people and property. Article 415 of the Civil Code provides, in general terms, that "whoever by his fault caused damage to another person shall be obliged to redress it". Generally speaking, a tort is established under Polish law if: (i) damage has occurred; (ii) an unlawful act was perpetrated; (iii) there is a causal link between the act and the damage; and (iv) the perpetrator of the act is at fault. According to Article 363 of the Civil Code, damage is remedied, at the election of the injured party, either by the previous condition being restored or a relevant sum of money being paid. If, however, restoration of the previous condition is not possible or it results in excessive difficulties or costs for the perpetrator of the damage, the injured party's claim must be limited to monetary performance. If the damage is to be remedied by payment of money, the value of the compensation is calculated as the date of determination of compensation, unless extraordinary circumstances require that value as at another date be used.

According to Article 444 of the Civil Code, in case of a bodily injury or a disturbance to health, any financial redress shall cover all resulting costs. The injured party can request the wrongdoer to deposit in advance the sum needed to cover the costs of medical treatment, and if the injured person has been rendered unable to continue in his present occupation, the sum needed to cover the costs of retraining for an alternative occupation. If the injured person has completely or partially lost his ability to be employed, or if the injured person's needs have increased or the chances of future success decreased, the injured person may claim an appropriate pension from the liable party. In addition, the court may award an appropriate sum to the injured person as a pecuniary compensation for the wrong suffered (Article 445, Section 1 of the Civil Code).

The settlement proceedings set out by the Geological and Mining Law must first be exhausted before a party is able to pursue a court claim for damage resulting from a breach of the law.

3.2 State regulatory action

Each European state has its own laws and regulatory regimes relevant to all aspects of the shale gas development process and which will give rise to regulatory and enforcement actions. Some of these laws are unique to each state, but many derive from regulations and directives of the European Union, which all EU member-states are obliged to implement into their own laws.

Although the European Union has until now stated that it does not intend to introduce any specific legislation relating to shale gas development, including, in particular, to address fracking, there are many existing EU regulations and directives that will directly and indirectly impact on shale gas developers and give rise to regulatory actions, including, for example: the Environmental Liability Directive, the Groundwater Directives, the Habitats Directive and the Air Quality Directive.

Touching briefly on the Environmental Liability Directive (2004/35/EC), this directive (which has been implemented in England through the Environmental Damage (Prevention and Remediation) Regulations 2009, and in Poland through the Prevention and Remedying of Environmental Damage Act of 13 April 2007) seeks to establish a common European framework for the prevention and remedying of environmental damage.

The provisions of the directive are implemented in different ways in each EU member state. However, as a minimum the directive requires member states to impose a positive obligation on an operator that is carrying out an activity that may give rise to environmental damage (of which shale gas development would be one example) to take necessary measures to prevent damage and to remedy any damage that has occurred. The directive also imposes a positive obligation on the operator to report either imminent threats of damage or occurrences of damage to the member state's designated competent authority. Under the directive, the competent authority will be able to require the operator to take remedial action, the costs of which will be borne by the operator. Finally, the directive provides that people who are affected by environmental damage, or have an interest in any administrative decision taken in relation to the damage, are entitled to request the competent authority to take action under the directive.

At the time of writing, the impact of the United Kingdom's decision to leave the European Union ('Brexit') on relevant regulations introduced into its laws as a result of its membership of the European Union is unclear. There will be various options open to the UK government in this regard: retaining the existing laws derived from EU regulations; introducing a bespoke new system of regulations to deal with the same matters; or a mixture of the two.

The authors would like to thank Hogan Lovells associates Brenna Finn and Aaron O'Connell, Hogan Lovells research analyst Margery Fernald, Hogan Lovells paralegal Danielle Hartley, and Hogan Lovells legal assistant Helen Hyatt for their immeasurably beneficial assistance with this chapter.

Shale gas regulation in the United Kingdom*

Yanal Abul Failat
LXL LLP

1. Introduction

The UK Government highly supports the trajectory of shale gas development in the United Kingdom. In its Gas Generation Strategy Paper published on 5 December 2013, it recognised that shale gas production would decrease reliance on imports and thus enhance the United Kingdom's energy security.[1] Moreover, the UK Institute of Directors report on 'UK Shale Gas Potential' explains that in the United Kingdom, there is the potential for production peaking at approximately 1.13 trillion cubic feet (tcf); a sector that could support around 70,000 jobs and security in respect of net benefit to the Treasury in tax revenues. On this basis, there has been a growing interest in the benefits of exploring the United Kingdom's shale gas, but a combination of technical challenges faced in shale gas operations, stern opposition from environmentalists, and concerns about the adequacy of the legal framework has slowed the progress of the emerging UK shale industry.

The United Kingdom's newfound interest in the shale industry and in employing hydraulic fracturing is not unprecedented.[2] Onshore exploration and production can be traced back to circa 1850; and since the 1950s, approximately 200 wells have been hydraulically fractured both onshore and offshore.[3] The first onshore hydraulic operations for extracting shale gas in the United Kingdom were conducted in 1991 and 1992 in West Sussex and Cheshire respectively.[4] Interestingly enough, the operations were so indigenous to the United Kingdom that the main active component in the fracturing fluid used was Marmite![5] Nonetheless, the fact is that until recently, exploring for shale gas was not financially viable and conventional resources were preferred as they require simpler technology and cost less to produce.

This inclination is because shales differ from conventional sources in that they are organic-rich, fine-grained sedimentary rocks, and they are both the source of and

* This chapter is based on Yanal Abul Failat, 'Fracking the UK's Shale Gas Regulatory Regime' (2014) 12 OGEL 3 www.ogel.org/article.asp?key=3472 accessed 30 August 2016.

1 Department of Energy and Climate Change, 'Gas Generation Strategy' (Presented to Parliament by the secretary of state for energy and climate change, Cm 8407, 2012) https://www.gov.uk/government/uploads/system/uploads/attachment_data/file/65654/7165-gas-generation-strategy.pdf accessed 21 January 2014.

2 UK Onshore Operators Group, 'UK Onshore Extraction – History' www.ukoog.org.uk/elements/pdfs/UKOOG%20Onshore%20Oil%20and%20Gas%20in%20the%20UK.pdf accessed 22 January 2014.

3 *Ibid.*

4 Shale Gas Europe, 'The UK's Oil and Gas Industry' (Shale Gas Europe, 24 February 2015) http://shalegas-europe.eu/the-uks-oil-and-gas-industry/?lang=de accessed 20 September 2016.

5 Michael Quentin Morton, 'Unlocking the Earth: a Short History of Hydraulic Fracturing' (2013) 10 (6) Geo ExPro 86, 90.

the reservoir for natural gas.[6] Although shale gas, like natural gas, is mostly composed of methane, it requires sophisticated technologies to drill and stimulate the gas-bearing zones as it is tightly locked in small spaces within the reservoir rock.[7] Shales are less porous than other types of rock, and their pores are so small as to render them relatively impermeable to gas flow.[8] In the United States, however, advanced technologies were employed to increase permeability mechanically, making the production of shale gas commercially viable. The process involves drilling a borehole into the shale horizon once it has been explored.[9] For a borehole to access a larger volume of rock, it is drilled horizontally.[10] The drilling is followed by, or rather combined with, hydraulic fracturing ('fracking'), which entails pumping water or liquid at high pressure into the rock and so inducing fractures increasing the permeability of the rock.[11] The geology and mechanical properties of rocks, stress fields and the employment of applicable stimulation procedures further complicate horizontal drilling, but fracking is what has made the production of shale gas possible.[12]

One look at the global energy sector is sufficient to see that due to this technology we are entering into a revolutionary aeon of unconventional hydrocarbon exploration and production. The booming US shale production has changed the international energy market and set an example for other countries to learn from and follow. In fact, the United States is not the only major player in the rising shale market; countries such as Argentina, Canada, China and Russia have a leading role not only in production, but also in developing technical expertise. The industry has globalised and many other countries – particularly countries that lack conventional resources, such as Chile, Jordan and South Africa – are happy to explore the prospects of shale gas production or other unconventional hydrocarbons to benefit their economies and achieve energy security.[13] As the *Petroleum Economist* puts it, "only a brave man would now bet against the emergence of a truly global shale-gas industry."[14]

1.1 The reserves

In the United Kingdom, the prospectivity of shale gas from shale rock is yet to be known. The Parliamentary Office of Science and Technology states that "there are currently no official reserve estimates... the UK reserves could be anywhere from zero to substantial."[15] A variety of estimates have been published. In 2010, the British Geological Survey estimated that the UK potential reserves are approximately

6 Gene Whitney and Carl E Behrens, *Energy: Natural Gas: the Production and Use of Natural Gas, Natural Gas Imports and Exports, Epact Project, Liquefied Natural Gas (LNG) Import Terminals and Infrastructure Security* (The Capitol Net Inc 2010) 2.

7 Jayarama Reddy, *Clean Coal Technologies for Power Generation* (CRC Press 2013) 51.

8 Whitney (n 6).

9 James Beard, *Environmental Chemistry in Society, Second Edition* (CRC Press 2013) 125.

10 British Geological Survey, 'New Shale Gas Resource Figure for Central Britain' (2014) https://www.bgs.ac.uk/shalegas/#ad-image-0 accessed 21 January 2014.

11 Beard (n 9).

12 John Breyer, *Shale Reservoirs: Giant Resources for the 21st Century, AAPG Memoir 97* (AAPG 2012) 71.

13 PricewaterhouseCooper, 'Energy Transformation: The Impact on the Power Sector Business Model (13th PwC Annual Global Power and Utilities Survey, 2013) at 12 www.pwc.com/gx/en/utilities/global-power-and-utilities-survey/assets/pwc-global-survey-new.pdf accessed 22 January 2014.

14 *Petroleum Economist*, 'The Slow Rise of Global Unconventional Gas' (*Petroleum Economist*, 14 January 2014) www.petroleum-economist.com/Article/3297089/News-Analysis-Unconventional/The-slow-rise-of-global-unconventional-gas.html accessed 21 January 2014.

5.3tcf.[16] In 2011, Cuadrilla Resources estimated an approximate 20tcf of recoverable shale gas in the UK basins of Bowland and West Lancashire.[17] In 2013, the US Energy Information Administration estimated that the UK holds 26tcf of shale gas and 26 million barrels of shale oil, reflecting the prospective resources within Carboniferous and Jurassic-age shale formations.

Figure 1: British onshore oil and gas production[18]

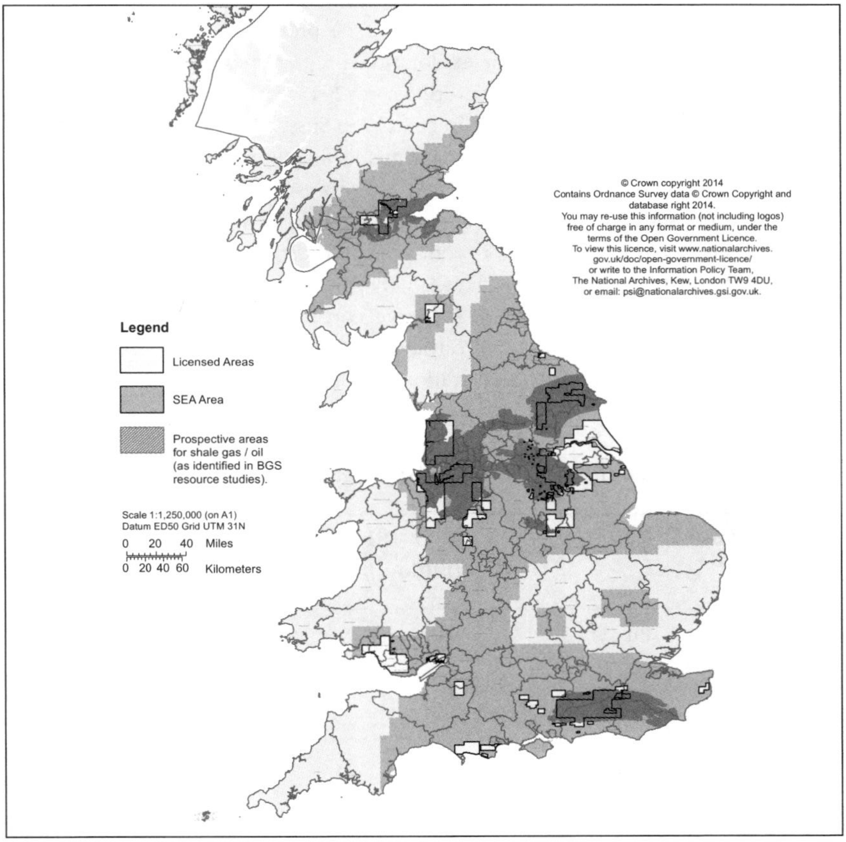

Source: Department of Energy and Climate Change. Contains public sector information licensed under the Open Government Licence v.3.0 available at: www.nationalarchives.gov.uk /doc/open-government-licence/version/3/

---

15 Parliamentary Office of Science and Technology, 'UK Shale Gas Potential' (Houses of Parliament, July 2013) www.parliament.uk/documents/post/ShaleGas_POSTbox.pdf accessed on 24 January 2014.

16 Department of Energy and Climate Change, 'The Unconventional Hydrocarbon Resources of Britain's Onshore Basins – Shale Gas' (2012) at 10 https://www.og.decc.gov.uk/UKpromote/onshore_paper/UK_onshore_shalegas.pdf accessed 24 January 2014.

17 Department of Energy and Climate Change (n 1) 54.

18 The map is available at https://www.gov.uk/government/publications/map-of-onshore-licences-sea-areas-and-prospective-areas. Permission to use the map is granted under the Open Government Licence. Please find the licence on www.nationalarchives.gov.uk/doc/open-government-licence/version/3/.

However, these calculations do not take into account any risked reserves, nor do they stipulate the proportion that can be economically produced. This is of particular importance when taking into consideration that an appropriate estimate of recoverable gas requires several wells to be drilled followed by testing of production characteristics. Production properties differ by location and rock formation, and thus resource estimates "have been subject to frequent and dramatic changes".[19] Only the profitability of the shale gas, however, can determine the true volume of recoverable shale gas in the United Kingdom. Howard Rogers of the Oxford Institute of Energy Studies explains that determination of the profitability of UK shale gas may well involve a two to three-year exploration programme, and that realistically it may require 50 to 100 wells to be drilled. In 2013, both Cuadrilla Resources and IGas conducted test drilling in Balcombe in West Sussex and the North West (the Bowland Shale) respectively.[20] The latter estimated an approximate volume of 102tcf (2.9 trillion cubic metres (tcm)) and updated this estimate in 2014 to 80tcf (2.3tcm).[21]

The reserves, however, are promising; the Bowland Shale's thickness, for instance, shows potential for profit even at a cost base initially expected to be higher than in the United States.[22] This is why in countries where the shale sector is more developed, estimates tend to be more consistent and accurate. The world's total technically recoverable shale oil and shale gas resources are estimated by the US Energy Information Administration to be 418.90 billion barrels and 7,576.6tcf respectively.[23]

Figure 2: Technically recoverable shale gas estimates

Rank	Country	Tight Oil (Billion Barrels)	Country	Wet Shale Gas (TCF)
1	Russia	74.6	China	1115.2
2	United States	78.2	Argentina	801.5
3	China	32.2	Algeria	706.9

continued on next page

19 *Ibid*, 55.

20 Jeanne Delebarre, Elena Ares and Louise Smith, 'Shale Gas and Fracking' (House of Commons Briefing Paper, Number 6073, 4 January 2017) at 9 http://researchbriefings.files.parliament.uk/documents/SN06073/SN06073.pdf accessed 14 January 2017.

21 IGas, 'Shale Gas in place in IGas' North West Licences of up to ca.170 tcf' (Euro Investor, 3 June 2013) http://ir1.euroinvestor.com/asp/ir/IGas/NewsRead.aspx?storyid=12357614&ishtml=1 accessed 15 January 2017; IGas, 'Preliminary Results for Nine Months to 31 December 2015' (IGas, 31 December 2015) www.igasplc.com/media/30438/final-results-presentation.pdf accessed 14 January 2017.

22 Department of Energy and Climate Change (n 1).

23 Technically recoverable "represents the volumes of oil and natural gas that could be produced with current technology, regardless of oil and natural gas prices and production costs"; US Energy Information Administration, 'Technically Recoverable Shale Oil and Shale Gas Resources: An Assessment of 137 Shale Formations in 41 Countries outside the United States' (EIA, 2015) at 10 https://www.eia.gov/analysis/studies/worldshalegas/ accessed 15 January 2017.

Rank	Country	Tight Oil (Billion Barrels)	Country	Wet Shale Gas (TCF)
4	Argentina	27.0	United States	622.5
5	Libya	26.1	Canada	572.9
6	Australia	15.6	Mexico	545.2
7	Venezuela	13.4	Australia	429.3
8	Mexico	13.1	South Africa	389.7
9	Pakistan	9.1	Russia	284.5
10	Canada	8.8	Brazil	244.9
World Total		**418.9**		**7,576.6**

Source: US Energy Information Administration. The table contains statistics provided by the US EIA, which is cited. No consent necessary.

1.2 The moratorium

When looking at the media, it is clear that there is international and national discourse on the nature of exploitation of shale formations, its effects on the environment and its regulation. When the United Kingdom expressed its interest in shale gas production, a strong opposition to fracking brought this resource into focus. Antagonism was triggered by YouTube videos of flaming taps, scrutiny by environmentalists and, of course, the induced seismicity experienced in Lancashire in 2011.[24] The British Geological Survey has found that the seismicity was the result of shale gas fracking in Poulton-le-Fylde operated by Cuadrilla Resources. The consequence was a temporary suspension of all shale gas fracking operations in the United Kingdom until the moratorium was lifted by Edward Davey in his ministerial statement to Parliament published by the Department of Energy and Climate Change on 13 December 2012.[25] In the ministerial statement, it was stated that after careful review of the scientific and engineering evidence on shale gas extraction conducted by the Royal Society and the Royal Academy of Engineering, the Department of Energy and Climate Change indicated that the available controls could mitigate the potential risks associated with such production.[26] The Department

24 Mark Brierley, 'The Future of Shale Gas in the UK' (*NewStatesman*, 13 January 2014) www.newstatesman.com/business/2014/01/future-shale-gas-uk accessed 21 January 2014; British Geological Survey (n 10).

25 Department of Energy and Climate Change, 'Written Statement by Edward Davey: Exploration for Shale Gas' (Edward Davey's Written Statement to Parliament, 13 December 2012) https://www.gov.uk/government/speeches/written-ministerial-statement-by-edward-davey-exploration-for-shale-gas accessed 21 January 2014.

26 *Ibid.*

of Health supported the Department of Energy and Climate Change's view and explained that potential risks to public health related to shale gas activities are rather low and that any such risks arise from operational failures and/or negligent on-site management.[27] It was further stated that "the risks can be managed effectively if operational best practices are implanted and enforced through regulations."[28] All the same, in July 2016, the Committee on Climate Change concluded that there is uncertainty surrounding the implications of shale gas in respect of greenhouse emissions, and that significant shale gas exploitation may be incompatible with UK carbon budgets unless three tests are met, which regulate:

- the limit of production emissions;
- shale gas consumption (ie, shale gas production should substitute imported gas rather than add gas consumption); and
- carbon reduction, including finding abatement measures compensating emissions arising from production.[29]

Four years on and shale gas exploration has developed differently within the United Kingdom. In Scotland, under the recent Scotland Act 2016, shale gas licensing was devolved to the Scottish Parliament,[30] which announced a moratorium on fracking in Scotland on 28 January 2015.[31] Wales is following suit; under the Wales Bill 2016/2017, provisions are included to devolve shale gas licensing to the National Assembly for Wales,[32] which issued a Direction on 13 February 2015 preventing local planning authorities from approving developments including fracking.[33]

In England, however, the shale gas sector has already embarked on an exploratory phase. While no commercial production operations have been authorised, the recent approval of the following two planning decisions indicates the possibility of the sector progressing into production in the future:

- In June 2015, Lancashire County Council rejected a planning application by Cuadrilla Resources for exploration at two sites at Preston New Road and Roseacre in Lancashire. Cuadrilla Resources made an appeal to the secretary of state for communities and local government, who in October 2016 permitted the application to progress to the next stage.[34]

---

27 Executive Agency of the Department of Health, 'Review of the Potential Public Health Impacts of Exposures to Chemical and Radioactive Pollutants as a Result of Shale Gas Extraction' (Draft for Comment, PHE-CRCE-002, 30 October 2013 www.hpa.org.uk/webc/HPAwebFile/HPAweb_C/1317140158707 accessed 23 January 2014.

28 Royal Society and the Royal Academy of Engineering, 'Shale Gas Extraction in the UK: a review of hydraulic fracturing' (2012) at 4 https://royalsociety.org/~/media/Royal_Society_Content/policy/projects/shale-gas/2012-06-28-Shale-gas.pdf accessed 18 March 2014.

29 Committee on Climate Change, 'Onshore Petroleum: The Compatibility of UK Onshore Petroleum with meeting the UK's Carbon Budgets' (March 2016) https://www.theccc.org.uk/wp-content/uploads/2016/07/CCC-Compatibility-of-onshore-petroleum-with-meeting-UK-carbon-budgets.pdf accessed 4 January 2017.

30 Scotland Act 2016, Explanatory Notes.

31 Scottish Parliament, 'Moratorium called on fracking' (Scottish Government, 28 January 2015) http://news.gov.scot/news/moratorium-called-on-fracking accessed 8 January 2017.

32 Wales 2016–2017 Bill, available at http://services.parliament.uk/bills/2016-17/wales.html accessed 20 December 2016.

33 The Town and Country Planning (Notification) (Unconventional Oil and Gas) (Wales) Direction 2015.

34 Delebarre, Ares and Smith (n 19), 4.

- In May 2016, the first planning approval for fracking was granted to Third Energy at Kirby Misperton, North Yorkshire. Anti-fracking campaigners brought a judicial review application of the application; in December 2016, the court ruled that the planning approval is legal.[35]

2. The legal framework

At present, in the United Kingdom there is no legislation dedicated to the exploration and production of shale gas, but the existing extensive onshore oil and gas regime is being used to regulate such activities. The government lifted the moratorium on fracking on the basis that the current system is robust enough to ensure the industry's application of good practice throughout its exploratory operations.[36] According to the Department of Energy and Climate Change:

> *So far as the UK is concerned... the industry has a good record, and... there are already in place robust regulatory controls on all oil and gas activities... the existing regulatory framework already provides the means to ensure that the industry does apply good practice throughout its operations; and that it will do so consistently. But we are taking further steps to reinforce the regime.*[37]

The United Kingdom's regulatory framework is highly regarded internationally, but in the context of shale gas exploration it is found to be complex. The industry is concerned about its efficiency and applicability in practice as the planning and permitting responsibilities are divided among various regulatory institutions, including:

- the Oil and Gas Authority;
- the Department of Energy and Climate Change;
- the Office of Unconventional Gas and Oil;
- the Environmental Regulator;
- the Mineral Planning Authority;
- the Health and Safety Executive;
- the Coal Authority;
- the UK Onshore Operators Group (an industry body); and
- the British Geological Survey.

The government has published a roadmap outlining the process that must be undertaken before operators commence exploration and production of shale gas.[38]

2.1 The licence

The Petroleum Act 1998 vests all rights to the petroleum resources in Great Britain in the Crown.[39] An operator that intends to explore and produce shale gas must first obtain a landward production licence from the Oil and Gas Authority, which is

---

35 *Ibid.*
36 Department of Energy and Climate Change (n 24).
37 *Ibid.*
38 UK Onshore Operators Group, 'Fact Sheet: Onshore Oil and Gas Regulation' (2014) www.ukoog.org.uk/images/ukoog/pdfs/fact%20sheets/regulation.pdf accessed 11 February 2014.
39 Petroleum Act 1998, s 2(1).

Figure 3: Planning and permitting roadmap[40]

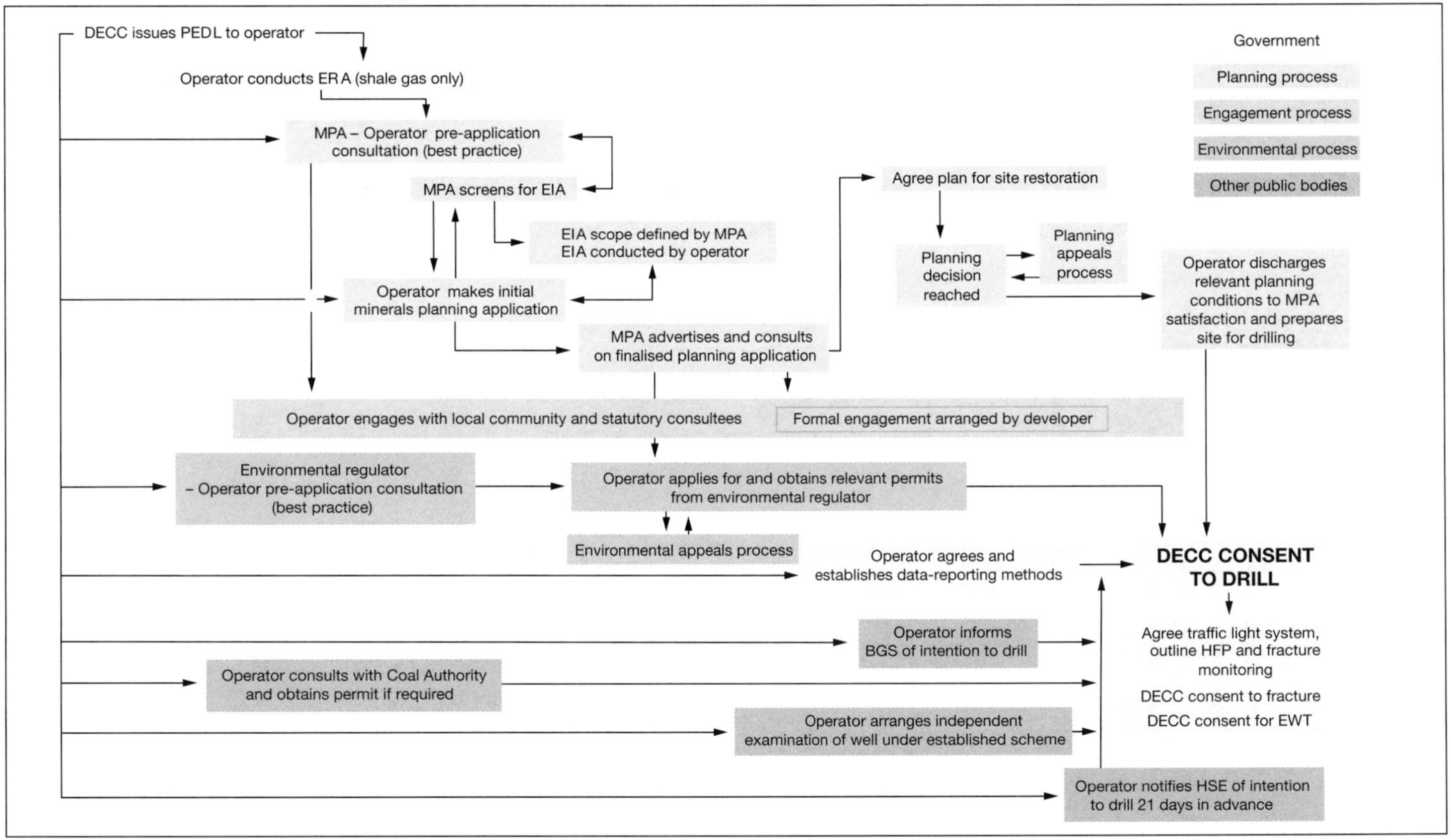

Source: Department of Energy and Climate Change. Contains public sector information licensed under the Open Government Licence v.3.0 available at: www.nationalarchives.gov.uk/doc/open-government-licence/version/3/. Permission to use the roadmap is granted under the Open Government Licence. Please find the licence on www.nationalarchives.gov.uk/doc/open-government-licence/version/3/.

empowered to grant such licences and required to make regulations that prescribe model clauses, which shall, unless the secretary of state sees fit to modify or exclude them in any particular case, be incorporated in any such licence.[41] Landward production licences typically entail an initial exploration term of five years, a second development term of five years, and a third production term of 20 years that brings about mandatory relinquishment of 50% of the licensed acreage at its end.[42]

The licence is designed to protect the licensee's investment from rivals that may seek to exploit hydrocarbons that are discovered, but also to ensure that the licensees "use it productively and do not simply bank it as a valuable but unused asset".[43] The landward model clauses (as set out in regulations enacted under the Petroleum Act 1998), which are incorporated into the terms of landward production licences, take into account the technologies and economics involved in shale gas production. For example, shale gas companies are unlikely to be able to surrender any acreage from the licence without any loss as shale gas is usually dispersed across the whole licence area, rendering the traditional surrender provisions inappropriate.[44]

The Oil and Gas Authority normally offers licences in competitive licensing rounds, which attract applications from single companies or a group of companies (commonly in the form of an unincorporated joint venture). In assessing a request for a licence and operatorship, the Oil and Gas Authority takes into consideration the operator's technical competency, financial standing, and operational and environmental management systems, as well as any insurance coverage available.[45] In the 14th licensing round (held and closed in 2015), which covered exploration for both conventional and unconventional oil and gas, 159 blocks were awarded.[46]

Landward production licences are not unique to shale gas and do not grant permission to drill, but rather provide the licensee with exclusivity on 'naturally occurring hydrocarbons', whether conventional or unconventional within a particular acreage. They also do not grant licensees any exemption from other regulatory requirements to conduct shale gas exploration and production activities, and operators must ensure that they obtain access rights from landowners and a chain of permits and permissions from governmental institutions that are designed to reinforce the existing legal framework before the Oil and Gas Authority grants final well consent.

---

40 The regulatory roadmap is available at https://www.gov.uk/government/publications/regulatory-roadmap-onshore-oil-and-gas-exploration-in-the-uk-regulation-and-best-practice.

41 Petroleum Act 1998, s 3(1) and s 4(1); The Petroleum Licensing (Exploration and Production) (Landward Areas Regulations 2014 are incorporated into the landward production licence).

42 Department of Energy and Climate Change, 'Oil and Gas: Petroleum Licensing Guidance' (Government Official Site, 4 December 2012) https://www.gov.uk/oil-and-gas-petroleum-licensing-guidance#types-of-licence accessed 10 February 2014.

43 Petroleum Licensing (Exploration and Production) (Landward Areas) Regulations 2014, Explanatory Memorandum.

44 *Ibid.*

45 Oil and Gas Authority, 'Onshore Licensing' https://www.ogauthority.co.uk/licensing-consents/onshore-licences/ accessed 29 October 2016.

46 The Oil and Gas Authority maintains an interactive map in which it maps out all the licence blocks. The map is available at: https://decc-edu.maps.arcgis.com/apps/webappviewer/index.html?id=29c31fa4b00248418e545d222e57ddaa; Department of Energy and Climate Change, 'Preparing for the 14th Round of Onshore Oil and Gas Licences' 18 December 2013) http://webarchive.nationalarchives.gov.uk/20140403094322/http://blog.decc.gov.uk/2013/12/17/preparing-for-the-14th-round-of-onshore-oil-and-gas-licences/ accessed 21 January 2017.

An oil company that wishes only to search for petroleum and does not require an exclusive right to drill or produce can apply for an exploration licence. This is attractive to seismic contractors that wish to collect data rather than exploit resources themselves (as well as landward production licence holders that want to explore outside areas where they hold exclusive production rights). The licence grants a non-exclusive right to explore (not produce), covering all acreage outside areas that existing production licences cover (for acreage covered by a landward production licence, permission is needed from the licence holder). Landward exploration licences cover exploration for petroleum only, unlike offshore exploration licences, which cover non-intrusive exploration for hydrocarbon production, gas storage, carbon capture and sequestration, or any combination of them.[47] Exploration licences are granted for a three-year term, renewable for a further period of three years subject to one month's written notice.[48] Exploration may begin with seismic investigations to identify prospective structures. Most mineral planning authorities deem such exploratory work as permitted development.[49] Operators, however, must inform the Highways Agency if a survey is undertaken on highways. Further, operators are required to submit a Petroleum Operation Notice Form (14b: Notification of intention to carry out onshore (landward) geophysical surveys), which are to be assessed by the relevant mineral planning authority and statutory agencies.[50]

As exploration for shale gas is at its early stages in the United Kingdom, the term 'shale gas well' is still not well defined. Even so, below is a list of wells that have been drilled and were planned to drill into or through strata regarded as prospective for shale gas despite the primary purpose of the well.[51]

The above licence holders are mostly small explorers that have been leading the sector. Nevertheless, an interest is maturing among the international oil majors (eg, Total), and other companies are in the process of drilling or making plans to do so on other sites in order to explore the potential of shale gas. The Oil and Gas Authority confirms that, to date, the only application which it, or its predecessor the Department of Energy and Climate Change, has received for consent to perform hydraulic fracturing for shale gas is from Cuadrilla Resources Ltd for the Preese Hall 1 well; shale fracturing in respect of this well took place between 26 March and 31 May 2011.[52]

From the table above, it is only Cuadrilla Resources that has used fracking for shale gas on the Preese Hall site in Lancashire; the Department of Energy and Climate Change confirms that as of 18 February 2014, it has not received any further applications to frack for shale gas at that stage.[53] Poyry expects a stable yet small

47 Oil and Gas Authority, 'Offshore Licences' https://www.ogauthority.co.uk/licensing-consents/offshore-licences/ accessed 30 October 2016.

48 Landward production licence Model Clauses 4 and 5.

49 Department of Energy and Climate Change, 'Onshore Oil and Gas Exploration in the UK: Regulation and Best Practice' (December 2015) https://www.gov.uk/government/uploads/system/uploads/ attachment_data/file/503067/Onshore_UK_oil_and_gas_exploration_England_Dec15.pdf accessed 16 January 2017.

50 Oil and Gas Authority, 'PON 14B' available at: https://www.ogauthority.co.uk/exploration-production/petroleum-operations-notices/pon-14b/ accessed 13 January 2017.

51 Email from Department of Energy and Climate Change to Yanal Abul Failat on 19 February 2014.

52 Letter from the Oil and Gas Authority to Yanal Abul Failat dated 13 February 2017.

53 *Ibid.*

Figure 4: Shale gas wells

<table>
<tr><th>Shale Gas Well (Licence)</th><th>Well Name (Location)</th><th>Operator</th><th>Beneficiary (Group)</th><th>Percentage Interest</th></tr>
<tr><td>LJ/21-9 (PEDL 184)</td><td>Ellesmere Port 1 (Cheshire West and Chester)</td><td>Island Gas Limited</td><td>1. INEOS Upstream Limited (INEOS Industries)
2. Island Gas Limited (IGAS Plc)</td><td>1. 50%
2. 50%</td></tr>
<tr><td>LJ/06-7 (PEDL 165)</td><td>Anna's Road 1 (Lancashire)</td><td rowspan="6">Cuadrilla Resources Limited</td><td rowspan="6">1. Bowland Resources (No 2) Limited (Centrica Group)
2. Bowland Resources Limited (Centrica Group)
3. Cuadrilla Bowland Limited (Cuadrilla Resources)
4. Lucas Bowland (No 2) Limited (AJ Lucas Group Ltd)
5. Lucas Bowland (UK) Limited (AJ Lucas Group Ltd)</td><td rowspan="6">1. 6.25%
2. 18.75%
3. 51.25%
4. 13.75%
5. 10%</td></tr>
<tr><td>LJ/06-6Z (PEDL 165)</td><td>Becconsall 1Z (Lancashire)</td></tr>
<tr><td>LJ/06-6 (PEDL 165)</td><td>Becconsall 1 (Lancashire)</td></tr>
<tr><td>LJ/01-1Z (PEDL 165)</td><td>Grange Hill 1Z (Lancashire)</td></tr>
<tr><td>LJ/01-1 (PEDL 165)</td><td>Grange Hill 1 (Lancashire)</td></tr>
<tr><td>LJ/06-5 (PEDL 165)</td><td>Preese Hall 1 (Lancashire)</td></tr>
</table>

Source: Oil and Gas Authority

number of companies applying for licences annually until 2019, followed by a rapid growth during the production era where it is expected that 100 wells will need permitting annually thereafter until 2024.[54]

2.2 The landowners

If the drilling of a well entails horizontal drilling that will cross the boundary of private property, the operator must ensure from the outset that consent from the landowner(s) is obtained; otherwise, the drilling may constitute a trespass. In *Star Energy Weald Basin Limited v Bocardo SA*,[55] the court unanimously decided that the owner's title extends down to the strata, between 800 feet and 2,800 feet, through which Star Energy's wells, casing and tubing pass unless there has been a conveyance, at common law or by statute to someone else:[56]

- an owner who holds the paper title has the *prima facie* right to possess not only the strata below but also the subsurface strata;[57] and
- that the right to search and bore for petroleum granted under a licence cannot bind a landowner who was not a party to the licence.[58]

However, under the Infrastructure Act 2015, investors benefit from a statutory right to use the deep-level land for exploiting petroleum or deep geothermal energy in England and Wales, which is at a depth of at least 300 metres below the surface within a landward area.[59] The exercise of such a right would require oil companies to make payments to landowners for sums decided by the Oil and Gas Authority.[60] The government acknowledges that any such compensation should be low because works should not materially interfere with the landowner's property.

If the licensee is seeking to conduct works at depths less than 300 metres below surface level, it should at first instance try to negotiate consent with the landowner(s). If unsuccessful, the licence holders can seek to acquire ancillary rights under the Petroleum Act 1998 and the Infrastructure Act 2015 if the landowner unreasonably refuses to reach an agreement or demands unreasonable terms.[61] The court may grant the licence holder all necessary ancillary rights and easements, including the right to access the land and sink boreholes to search and extract shale gas, if it is satisfied that it is expedient in the national interest.[62]

The calculation of such compensation is subject to the statute, but in the

---

54 Poyry, 'Macroeconomic Effect of European Shale Gas Production (November 2013) Where are we Now?' (2014) 7; available at: www.fraw.org.uk/files/extreme/poyry_2014.pdf accessed on 23 February 2014.

55 [2010] UKSC 35.

56 *Ibid*, paras 27-28.

57 *Ibid*, para 31.

58 *Ibid*, para 32. The latter point is supplemented by s 9(2) of the Petroleum Act 1998, which reads: "Nothing in this Act shall be construed as conferring, or as enabling the [Secretary of State] to confer, on any person, whether acting on behalf of His Majesty or not, any right which he does not enjoy apart from this Act to enter on or interfere with Land."

59 This is measured vertically above the point where works take place, excluding buildings, structures and water; See Infrastructure Act 2015, s 43 and the Petroleum Licensing (Exploration and Production) (Landward Areas) Regulations 2014, s 1 which defines 'landward area' as "an area which lies on the landward side of lines drawn in accordance with the provisions of Schedule 1."

60 Infrastructure Act, s 45(1).

61 Infrastructure Act 2015, s 7(1); Mines (Working Facilities and Support) Act 1966, s 3(2).

62 Petroleum Act 1998, s 7(3)(a); Mines (Working Facilities and Support) Act 1966, s 3(1).

Supreme Court case of *Bocardo*, the judges rejected the argument that the legal principles on compensation entitled the landowner to the equivalent of royalty in exchange for access rights.[63] When assessing the compensation, the court referred to two relevant provisions. Firstly, Section 8(2) of the Mines (Working Facilities) Act 1966, which provides that compensation would be assessed "on the basis of what would be fair and reasonable between a willing grantor and a willing grantee". Secondly, Section 3(2)(b) of the Petroleum Act 1934, which provides an additional allowance of at least 10% which shall be made upon a compulsory acquisition of rights; this is now mirrored in Section 7(4)(b) of the Petroleum Act 1998. The court also quoted Lord Nichols in saying that "the pressing need for an acquiring authority for land as part of a scheme was to be disregarded in assessing the land's value for the purposes of compensation"; this is referred to as the Pointe Gourde principle.[64] Accordingly, in the court of appeal, an award of damages of £1,000 was upheld.[65]

As the current framework stands, although landowners may only be entitled to nominal compensation, the reality is that this approach may render the development of projects impractical. The UK Onshore Operators Group estimated that in cases where an agreement is not reached between a developer and a landowner, the Petroleum Act 1998 compulsory process would take 18 to 24 months.[66] This period clearly does not mirror the process in the United States, where landowners own the minerals in their land and achieve higher financial or economic return when granting companies any rights of access and/or ownership. Further, the UK Onshore Operators Group's summary set out below has been viewed as time consuming and not viable for projects:

(a) *"Determine the expected well path, allowing for margins of error should the well deviate from its planned path;*

(b) *Identify the land ownership under which the well is expected to pass;*

(c) *Enter into negotiations with the landowner to (or "intending to") obtaining the necessary grant of rights to drill under their land;*

(d) *If the landowners are too numerous, refuse the grant or demands unreasonable terms, apply to the Minister ([OGA]) for ancillary rights;*

(e) *The Minister will review the case, including consultation with the landowners and, unless he considers that a prima facie case is not made out, refer the matter to the Court; and*

(f) *The Court will set a date and hear the case and, if it is satisfied that the requirements of the legislation are complied with in the case of the applicant licence holder, grant the compulsory purchase right."*[67]

If the investor and the landowner cannot agree, then the investor can seek to acquire ancillary rights under the Petroleum Act 1998. Before seeking ancillary

63 Petroleum Act 1998 s 7(4)(b); Terence Daintith, *Finders Keepers? How the Law of Capture Shaped the World Oil Industry* (Routledge 2010) 324.

64 *Waters v Welsh Development Agency* [2004] UKHL 19, paras 64-65.

65 *Bocardo* (n 53), paragraph 92.

66 Calum Stacey, 'Case Comment: *Bocardo SA v Star Energy UK Onshore Ltd and another*' (2011) 2 (1) *Aberdeen University Law Review* 124, 134.

67 *Ibid.*

rights, the investor must show that private negotiations between themselves and the landowners were unsuccessful because:

- there is a large number of landowners (who may well have conflicting interests);
- the title and/or the rights cannot be ascertained;
- the affected landowners do not have the right to dispose of their interest (ie, due to a defect in title or a legal disability); and
- the affected landowners unreasonably refused to grant the rights or demand terms which are unreasonable.

2.3 Environmental risk assessment

An environmental risk assessment (ERA) is a first-stage risk assessment to be conducted for proposed shale gas operations where fracking is planned. As a matter of good practice, the Oil and Gas Authority requires a compilation of an ERA as a matter of good practice to review before granting final consent for drilling. The relevant mineral planning authority determines the need for an ERA by undergoing a screening exercise under European law. Most or all proposals which involve fracking will require an ERA.[68] The ERA is recommended as a starting point for early engagement with the mineral planning authority, other regulators and communities, and therefore should be taken as soon as possible and, in any case, before applying for planning consent.

In its report of 2012, the Royal Society provides that such an assessment should be compulsory for all shale gas operations and must be assessed across all the stages of such project – namely from exploration to disposal.[69] At the moment, however, the ERA focuses on the exploration phase of shale gas activities and covers:

- baseline monitoring;
- water acquisition;
- chemical mixing;
- borehole integrity;
- well injection;
- flowback fluid management, including residual material left in the well;
- gas management;
- offsite disposal or reuse; and
- well decommissioning.[70]

Further, if it is shown that any European-protected species or habitats may be affected by shale gas exploration, an assessment under the Habitats Directive is also required;[71] an obligation which was implemented into national legislation. In the 14th onshore licensing round, 132 out of 159 awarded blocks were subject to

---

68 UK Onshore Operators Group, 'Who is Responsible for What?' (2014) www.ukoog.org.uk/knowledge-base/regulation/who-is-responsible-for-what accessed 11 February 2014.

69 Royal Society and Royal Society of Engineering (n 27) 5.

70 Environment Agency, 'An Environmental Risk Assessment for Shale Gas Exploratory Operations in England' (Version 1, August 2013) http://a0768b4a8a31e106d8b0-50dc802554eb38a24458b98ff72d550b.r19.cf3.rackcdn.com/LIT_8474_fbb1d4.pdf accessed 12 February 2014.

extensive assessment and a consultation on the approach to providing landward production licences in light of the Conservation of Habitats and Species Regulations 2010.[72] A procedure was proposed by the consultation for determining European-protected areas, which may be affected by licence activities areas and implementing policies on incorporating licence conditions that would protect such areas from fracking.[73]

2.4 Environmental impact assessment

In supplement to the results of the ERA, an environmental impact assessment (EIA) is usually required. An EIA is mandatory for projects which fall within the applicable thresholds and criteria of Schedules 2(d) and 3 (extractive industry) to the Town and Country Planning (Environmental Impact Assessment) Regulations 2011. The objective of EIAs is to establish the possible effects of developments involving fracking on the environment and to ensure that these effects are considered and mitigated before any shale gas project is allowed to commence.[74] Operators are advised to contact the mineral planning authority for an initial screening opinion to determine whether an EIA is required. If one is needed, then an environmental statement should be submitted to the mineral planning authority as part of the planning application process.[75] A planning permission for a proposal which is subject to an EIA and that has not been screened or is not accompanied by an EIA could be challenged.

Depending on the activities proposed, the EIA may include descriptions of the development, activities, type, quantity and location of potential emissions from the operations, soil and water contamination mitigation measures.[76] In turn, the authority may assess the impacts of the proposed operations on water, air quality, flooding, ecological receptors and other hydrogeological risks.[77] It should be noted that certain areas, including national parks, roads, world heritage sites and protected groundwater source areas, are protected at depths of less than 1,200 millimetres from the surface from fracking activities.[78]

An EIA is required only where a project is likely to have a significant environmental effect. While applications are assessed on a case-by-case basis, it is unlikely to be needed for exploratory drilling operations which do not involve fracking. An EIA is mandatory for applications where it is envisaged that more than 500 tonnes of oil or 500,000 cubic metres of gas respectively would be extracted per day.[79] Further, other applications for exploratory, appraisal and production phases

---

71 Council Directive 92/43/EEC.
72 Delebarre, Ares and Smith (n 19) 14.
73 *Ibid.*
74 Energy and Climate Change Committee, *Shale Gas: Fifth Report of Session 2010-2012* (Stationary Office 2011) 23.
75 *Ibid.*
76 Environment Agency, 'Onshore Oil and Gas Sector Guidance (Version 1, 17 August 2016) https://www.gov.uk/government/uploads/system/uploads/attachment_data/file/545924/LIT_10495.pdf accessed 29 November 2016.
77 *Ibid.*
78 Onshore Hydraulic Fracturing (Protected Areas) Regulations 2015.
79 Town and Country Planning (Environmental Impact Assessment) Regulations 2011, Schedule 1.

may require an EIA if they exceed the applicable threshold or any part of the development to be carried out in a sensitive area.[80] Applications are assessed on a case-by-case basis and it is unlikely that an EIA will be required for exploratory drilling operations which do not involve fracking.

2.5 Planning application

Before any drilling starts, planning permission is an approval that an operator must obtain from the relevant mineral planning authority; this is either the county council in England or the planning authority in Scotland and Wales.[81] The planning regime complements the other regulatory regimes, and as expressed in the National Planning Policy, it aims to assess the appropriateness of the activities for their location and any risks to health, natural environment and general amenity.[82] In evaluating an application for planning permission, economic, social and environmental factors such as noise associated with the operation, dust, air quality, traffic and other effects on the landscape will be taken into account.[83] Separate planning permissions are required for each phase of shale gas extraction. Nonetheless, the Town and Country Planning (General Permitted Development) Order 1995 provides that there may be a deemed consent that follows initial seismic work.[84] These applications, which are assessed on a case-by-case basis, will also be subject to the consultation of the Environment Agency for England and the local communities.[85]

The application and any other information that is material to it will be published and available for inspection by members of the public who are encouraged to engage and comment on the proposed activities.[86] When a decision is made on the planning application, only 'material considerations' can be taken into account. There is no published list of what is deemed a material consideration in this context, although the courts have excluded matters such as loss of property, loss of view and opposition to the principle of development representations as non-material.[87] Some light is shed by the National Planning Policy Guidance, which set outs some 'principal issues' for consideration – namely: noise associated with the operation; dust; air quality; lighting; visual impact on the local and wider landscape; landscape character; archaeological and heritage features; traffic; risk of contamination to land; soil resources; geological structure; impact on best and most versatile agricultural land; blast vibration; flood risk; land stability; wildlife sites; protected landscapes; and water abstraction.[88]

---

80 *Ibid*, Schedule 2.
81 Department of Energy and Climate Change (n 47).
82 Department for Communities and Local Government, 'Planning Practice Guidance for Onshore oil and Gas' (2013) at 7 https://www.gov.uk/government/uploads/system/uploads/attachment_data/file/224238/Planning_practice_guidance_for_onshore_oil_and_gas.pdf accessed 24 March 2014.
83 *Ibid.*
84 Town and Country Planning (General Permitted Development) Order 1995, Schedule 2.
85 Department of Energy and Climate Change (n 4779).
86 Department of Energy and Climate Change, 'Developing Onshore Shale Gas and Oil – Facts about Fracking' (2013) at 8 https://www.gov.uk/government/uploads/system/uploads/attachment_data/file/265972/Developing_Onshore_Shale_Gas_and_Oil__Facts_about_Fracking_131213.pdf accessed 4 March 2014.
87 *Frack Free Balcombe Residents v West Sussex County Council* [2014] EWHC 4108.
88 Department for Communities and Local Government, 'Minerals: what are the environmental issues of minerals working that should be addressed by mineral planning authorities?' (DCLG, Planning Practice Guidance, 17 October 2014) https://www.gov.uk/guidance/minerals accessed 15 January 2017.

2.6 Environmental permits

Onshore oil and gas exploratory activities, whether for conventional or unconventional hydrocarbons, require bespoke permits depending on the techniques used for exploration and the geology of the site.[89] These must be obtained from the relevant environmental regulator. As a first step in obtaining these permits, an operator that intends to "construct or extend a boring for the purpose of searching for or extracting minerals" must first serve a notice to the Environment Agency.[90] Then it must apply for the relevant environmental permits in accordance with the Environmental Permitting (England and Wales) Regulations 2010, including environmental permits for:

- a mining waste operation involving the "management of extractive waste, whether or not involving a mining waste facility".[91] Any mining waste activities forming part of shale gas exploration will require this environmental permit if the aim is to establish the existence, characteristics and quality of the hydrocarbons reserves of economic value;
- water discharge activity including the "discharge or entry to inland freshwaters, coastal waters or relevant territorial water of any (i) poisonous, noxious or polluting matter, (ii) waste matter, or (iii) trade effluent or sewage effluent";[92]
- groundwater activity or, in other words, "the discharge of a pollutant that results in the direct input of that pollutant to groundwater activity";[93]
- naturally occurring radioactive material industrial activity. Fracking for shale will involve the production of natural gas which is listed under the Environmental Permitting (England and Wales) Regulations 2010. Therefore, shale gas activities post the exploration phase will be subject to this permit;[94] and
- any flaring with a capacity of more than 10 tonnes of natural gas per day.[95]

Further, water abstraction from groundwater and surface water may trigger the need for further consents if the abstraction is anticipated to exceed 20 cubic metres (m^3) per day of water. In such a case, an application to the Environment Agency must be made to acquire an abstraction licence.[96] If the abstraction is expected to affect groundwater, then groundwater investigation consent is required to cover the construction and test pumping of a borehole before an application for an abstraction licence is made.[97] An operator may also be subject to flood defence consent if the relevant site for exploration activities is in, over or under a watercourse which is part of the main river.[98]

89 Environment Agency, 'Unconventional Gas Permits' (EA, 2014) www.environment-agency.gov.uk/business/topics/134508.aspx accessed 18 February 2014.
90 Water Resources Act 1991, s 199.
91 The Environmental Permitting (England and Wales) Regulations 2010, Schedule 20, paragraph 2(1).
92 The Environmental Permitting (England and Wales) Regulations 2010, Schedule 21.
93 *Ibid*, Schedule 22, paragraph 3.
94 *Ibid*, Schedule 23; Environment Agency, 'Guidance Note: regulation of Exploratory Shale Gas Operations' (2013) at 3 www.groundwateruk.org/downloads/EA_ShaleGasRegulation.pdf accessed 12 February 2014.
95 The Environmental Permitting (England and Wales) (Amendment) Regulations 2013, Regulation 35.
96 The Water Resources Act, s 24.
97 *Ibid*, s 32.
98 *Ibid*, s 109.

For oil and gas operations, applicants may choose to apply for a 'standard rules permit', a bespoke permit or a combination of both depending on the operations it intends to carry out. For oil and gas, there are standard rules available for:

- drill and core activities using water and oil-based muds;[99]
- accumulation and disposal of radioactive waste;[100]
- drill and core operations including leaking off testing, acid wash and decommissioning;[101] and
- storage and handling of crude oil.[102]

Licensees may apply for a bespoke permit if they cannot comply with the conditions set in those standard rules, are undertaking activities for which standard rules permits do not exist, or want to combine standard rules and bespoke permits. Where an operator wishes to apply for a combined licence for a single site, environmental guidance encourages a stated approach so that operators apply for permits in the order in which on-site activities may take place.[103] If a bespoke application is made, the Environment Agency will consult the public on the application, including the publication of the application details, arranging a consultation and receiving comments from the public, and liaising with other statutory agencies.

The closure of a well once a decommissioning is planned to take place must be in accordance with the Borehole Sites and Operations Regulations 1995, the Offshore Installations and Wells (Design and Construction, etc) Regulations 1996, the landward production licence and the Oil & Gas UK guidelines for the suspension and abandonment of wells.[104] The plugging and abandonment of a well process requires that:

- any permeable layers within the well are sealed;
- cement is filled into the remainder of the well;
- the wellhead is removed; and
- the casings are cut and sealed below ground level.

---

99 Environment Agency, 'SR2014 No 2 The management of waste, not including a waste facility, generated from onshore oil and gas prospecting activities without well stimulation (using water and/or oil based drilling mud)' (Standard Rules 9 December 2014) https://www.gov.uk/government/publications/sr2014-no-2-the-management-of-extractive-waste accessed 4 January 2017.

100 Environment Agency, 'SR2014 No 4 Accumulation and disposal of radioactive waste from the NORM Industrial Activity of the production of oil and gas' (Standard Rules, 9 December 2014) https://www.gov.uk/government/publications/sr2014-no-4-accumulation-and-disposal-of-radioactive-waste accessed 4 January 2017.

101 Environment Agency 'SR2015 No 1 The management of extractive waste, not including a waste facility, generated from onshore oil and gas prospecting activities including drilling, coring, leak off testing (LOT), acid wash and decommissioning, storage and handling of crude oil' (Standard Rules, 2015) https://www.gov.uk/government/uploads/system/uploads/attachment_data/file/504677/LIT_10411.pdf accessed 4 January 2017.

102 Environment Agency 'SR2015 No 2 Storage and handling of crude oil arising from onshore oil and gas exploration and production activities' (Standard Rules, 2015) https://www.gov.uk/government/uploads/system/uploads/attachment_data/file/504679/LIT_10414.pdf accessed 4 January 2017.

103 *Ibid.*

104 Oil & Gas UK, 'Guidelines for the suspension and abandonment of wells' (July 2012) http://oilandgasuk.co.uk/product/guidelines-for-the-suspension-and-abandonment-of-wells-issue-4-july-2012-and-guidelines-on-qualification-of-materials-for-the-suspension-and-abandonment-of-wells-issue-1-july-2012/ accessed 4 January 2017.

The main extractive waste that arises from well abandonment is surplus cement from well plugging, and broken concrete and cement from the removal of the well cellar and conductor.

Wells drilled before 1 October 2013 would not require an application for an environmental permit for a mining operation if the licence conditions have been complied with, the activities exclude the management of extractive waste involving a waste facility, and the extractive waste is managed without endangering or harming human health, environment (ie, water, soil, air, fauna and flora) and without causing nuisance through noise or odour, and without adversely affecting the landscape or places of special interest.[105]

2.7 The Health and Safety Executive notification

Operators that have plans to drill a shale gas well must, 21 days before any such plans,[106] notify the Health and Safety Executive, which monitors and promotes the adoption of safe working practices by operators. Its goal is to reduce risks to workers and the public as far as reasonably practicable via two specific regulations, namely the Borehole Sites and Operations Regulations 1995 and the Offshore Installations and Wells (Design and Construction, etc) Regulations 1996. The former mainly deals with the manner in which risks are managed on site, and the latter is more concerned with well integrity and control applying through the entire life cycle of shale gas wells. Both sets of regulations ensure that the operator reports to the Health and Safety Executive with the main information on a weekly basis throughout the various stages of the relevant activities so that it can properly assess the extent of risk control and mitigation taken by the operators.[107]

2.8 Well examination scheme

One of the features of the Offshore Installations and Wells (Design and Construction, etc) Regulations 1996 is that they require the operator to prepare and implement arrangements for an independent well examination that entails a complete examination of well design and construction.[108] Regulation 18 provides that the well examination scheme must be held by independent and competent persons who are also required to prepare and write reports and recommendations, with the assistance of the well operators, so that the well is designed, constructed and maintained in a condition that:

> *so far as is reasonably practicable, there can be no unplanned escape of fluids from the well; and risks to the health and safety of persons from it or anything in it, or in strata to which it is connected are as low as is reasonably practicable.*[109]

The well examination scheme must also consider well design aspects such as

---

105 Environment Agency, 'Onshore oil and gas well decommissioning and abandonment for wells drilled prior to 1 October 2016' (Regulatory Position Statement, January 2014) https://www.gov.uk/government/uploads/system/uploads/attachment_data/file/572304/LIT_9054_c96523.pdf accessed 24 May 2016.

106 Offshore Installations Safety Case Regulations 2005, Regulation 7(1).

107 Health and Safety Executive, 'Shale Gas' (HSE, 2014) www.hse.gov.uk/shale-gas/about.htm accessed 1 April 2014.

108 Offshore Installations and Wells (Design and Construction, etc) Regulations 1996, Regulation 18(1).

109 *Ibid*, Regulation 18(2).

groundwater and aquifer-isolation, fracture containment, induced seismicity risks, fracturing and flow-back testing programmes and operations.[110] The well examination scheme should also include examinations of decommissioning designs and processes.[111]

2.9 The Coal Authority permission

The Coal Authority, which was established by the Coal Industry Act 1994, has the right to "holding, managing and disposing of interests and rights in or in relation to the unworked coal".[112] Accordingly, an operator that intends to explore for shale gas in or through an area that intersects, disturbs or enters the Coal Authority's interest must request written permission from it.[113] The Coal Authority will mainly base its decision on the results of desktop studies undertaken to evaluate the impact of such activities on its interests.[114]

2.10 The well consent

When the chain of permits and actions are completed and obtained, the operator must inform the British Geological Survey of its intent to commence drilling and proceed to obtain the last well or fracking consent from the Oil and Gas Authority. The consent will be assessed and scrutinised on a case-by-case basis taking into account the operator's competency and financial stability.[115] More specifically, the Oil and Gas Authority will consider whether the operator has:

- proved that operational and environmental management systems are in place;
- secured planning permission from the mineral planning authority/local planning authority (and discharged conditions set out therein);
- obtained a permit from the Coal Authority (if required);
- informed the British Geological Survey of its intention to drill;
- completed consultation processes with all statutory consultees;
- obtained all necessary permits from and agreed on a system for monitoring conditions and emissions with the relevant environmental agency;
- provided the Health and Safety Executive with a minimum 21 days' notice of its intention to drill as well as details of the proposed well;
- agreed with the Oil and Gas Authority data reporting and monitoring induced seismicity; and
- received approval for its hydraulic fracturing programme.[116]

---

110 United Kingdom Onshore Operators Group, 'UK Onshore Shale Gas Well Guidelines' (1 February 2013) at 24 https://www.gov.uk/government/uploads/system/uploads/attachment_data/file/185935/UKOOGShaleGasWellGuidelines.pdf accessed February 28 2014.

111 *Ibid.*

112 The Coal Industry Act 1994, s 1.

113 Coal Authority, 'Permits for Intersecting Disturbing or Entering Coal' (2013) coal.decc.gov.uk/en/coal/cms/services/permits/permits.aspx accessed on 11 February 2014.

114 Coal Authority, 'Annual Report and Accounts 2012–2013 (2013) at 13 http://coal.decc.gov.uk/assets/coal/publicationsandinformation/transparency-foi/7607-coal-authority-annual-report-and-accounts-for-2012.pdf accessed 11 February 2014.

115 Department of Energy and Climate Change, 'Guidance on Fracking: Developing Shale Gas in the UK' (2017) https://www.gov.uk/government/publications/about-shale-gas-and-hydraulic-fracturing-fracking/developing-shale-oil-and-gas-in-the-uk accessed 13 January 2017.

116 UK Onshore Operators Group (n 66).

Further, the Infrastructure Act 2015 includes safeguards in respect of onshore fracking, including requiring independent well inspections, monitoring of groundwater, restoration conditions, banning fracking within protected areas and attaching the following requirements to well consents:

a) a condition which prohibits associated fracking from taking place at a depth of less than 1,000 metres; and

b) a condition which prohibits associated fracking from taking place in a land at a depth of 1,000 metres or more unless the secretary of state also grants hydraulic fracking consent.[117]

Operators that are planning to conduct fracking operations on a specific well have the option of bundling their requests for drilling and fracturing consents regardless of when they were sought (ie, before or after drilling of the well); the operator must obtain Oil and Gas Authority approval before fracturing commences.[118]

3. Timescales of planning and permitting process

The planning and permitting process has no defined timescales; in fact, the Oil and Gas Authority's regulatory roadmap expressly states that the "roadmap does not define timescales for the planning and permitting process or individual steps within it".[119] Indicative timescales are established by the relevant regulatory authorities. As per the Parliamentary Economic Affairs Committee, the table below highlights such indicative timescales.

Figure 5: Timescales for planning and permitting regime

Stage	Length of Time
Oil and Gas Authority	
Issue of petroleum exploration and development licence	Gained via licensing rounds
Environmental risk assessment	No indication was given

continued on next page

117 Infrastructure Act 2015, 50(1).
118 Department of Energy and Climate Change (n 47).
119 Parliamentary Economic Affairs Committee, 'The Economic Impact on UK Energy Policy of Shale Gas and Oil' (3rd Report of Session 2013/2014, 28 May 2014) at 73–75 www.publications.parliament.uk/pa/ld201314/ldselect/ldeconaf/172/172.pdf accessed 8 May 2014.

Stage	Length of Time
Local authorities	
Planning permissions	21-day consultation
Environmental impact assessment	13–16 weeks
Environment Agency	
Notice of intention to drill	Served, one month before drilling
Environmental permits	13–26 weeks
Health and Safety Executive	
Notice of intention to drill	Served 21 days before drilling

Source: House of Lords

4. Guidelines

The above statutory framework is supplemented by the UK Onshore Shale Gas Well Guidelines that were produced by a high-level workgroup that included operating and service companies, the Oil and Gas Authority (and the UK Onshore Operators Group), the Health and Safety Executive, the Environment Agency and the Scottish Environment Protection Agency.[120] The guidelines relate to aspects of exploration and appraisal, and contain industry best practices; guidelines for production and storage phases will follow. The guidelines are not mandatory, and a joint venture may adopt an alternative approach to managing well integrity, fracturing operations or environmental management. Conversely, adopting them will ensure that companies can demonstrate that they have adhered to all relevant regulations that apply to shale gas operations.[121] The guidelines are subdivided into the following:

- Well Integrity Guidelines;
- Guidelines for the Suspension and Abandonment of Wells;
- Guidelines on Qualification of Materials for the Suspension and Abandonment of Wells;
- Guidelines for Well Operators on Well Examination;
- Guidelines for Well Operators on Competency of Well Examiners; and
- Guidelines on Competency for Wells Personnel.[122]

Operators that are members of the UK Onshore Operators Group, the

120 UK Onshore Operators Group (n 108).
121 *Ibid.*
122 Stacey (n 64).

representative body of the UK onshore oil and gas industry, sign up to the UK Onshore Operators Group Community Charter, which promotes open and transparent communication between the industry, stakeholder groups and the relevant communities.[123] Among the charter's central objectives is ensuring that there is a continuous point of contact for local communities and engagement between them and the industry. Communities are crucial for the success of a shale gas company's drilling approvals process and can engage and influence the following: pre-application consultation; environmental risk assessment; pre-planning notices; planning authority consultation; the environmental impact assessment; and environmental permitting.[124]

Under the charter, UK Onshore Operators Group members commit to paying £100,000 to local communities that host their operations during the exploration stage and to providing a 1% share of proceeds at the production stage to the local community.[125] Moreover, operators are expected to publish an annual industry report to prove how the charter has been met and will be subject to regular review as the sector develops.[126]

5. Final remarks

In the United Kingdom, the exploration and production of shale gas is still in its formative years, but its continuous development is capable of changing the energy and environmental and socioeconomic landscape. On the premise that there are positive aspects of shale gas production, and that there is sufficient control to mitigate any risk, the government is eager to explore the sector to contribute to energy security, economic growth, job creation and tax revenues. The government is already committed to incentivising oil and gas companies to invest in the sector; a new tax allowance is being introduced for oil and gas companies, (operating in shale) by reducing the tax rate on a portion of a company's profits from 62% to 30% of 75% of their capital expenditure.[127] Moreover, the Parliamentary Economic Affairs Committee recommended that shale development should be recognised as a national priority and that a new committee of the Cabinet, chaired by the chancellor, should be established to ensure the United Kingdom's commitment to "go all out for shale".[128]

From a regulatory point of view, the statutory legal framework and the consenting regime certainly provide legislative coverage but not necessarily legal adequacy. The European Parliament has concluded that there are significant gaps in

123 UK Onshore Operators Group, 'Community Engagement Charter on Oil and Gas from Unconventional Reservoirs' (July 2013) https://www.gov.uk/government/uploads/system/uploads/attachment_data/file/256360/Publication_UKOOG_communityengagementcharterversion6.pdf accessed 15 April 2014; UK Onshore Operators Group, 'How to engage with Shale/Gas Hydraulic Fracturing Planning and Permitting' (13 January 2014) www.ukoog.org.uk/about-ukoog/press-releases/66-how-to-engage-with-shale-gas-hydraulic-fracturing-planning-and-permitting accessed 15 April 2014.

124 *Ibid.*

125 *Ibid.*

126 *Ibid.*

127 HM Treasury, 'Autumn Statement 2013' (2013) at 49 https://www.gov.uk/government/uploads/system/uploads/attachment_data/file/263942/35062_Autumn_Statement_2013.pdf accessed on 20 April 2014.

128 Parliamentary Economic Affairs Committee (n 117).

the legal framework despite the several legislative instruments that are relevant to fracking.[129] This is mainly due to the fast pace of development of fracking technologies, which the development of the law cannot necessarily keep up with. Moreover, there is a common consensus that the licensing and consenting regime is bureaucratic and time consuming; streamlining the process is fundamental to successful shale gas production.[130] The regime remains untested in the context of shale gas, and there are uncertainties as to the practicalities and timescales of the process.

In response, the regulatory bodies in the United Kingdom provide that the environmental risks for exploration have been thoroughly assessed against the existing legal framework, and that the regulation is robust and contains the right regulatory controls. Accordingly, the government already has plans to streamline the application process into a single application.[131] In fact, some would argue that the regulatory landscape in the United Kingdom is already more comprehensive than in the United States and the legislation sets high environmental thresholds.[132] Further, the permitting regime as it stands is only for approving the exploration phase; operators will need to apply for planning permission at each stage (appraisal and production), and further regulation for these steps will develop and follow.[133] Further, with the aid of light-handed regulation and best practices, via the UK Onshore Operators Group, the industry and the government can continue to cooperate and ensure that the regulation is developing alongside the technology. The industry is encouraging sensible regulation and acknowledging its importance due to the progressive technology and the lack of public confidence.[134]

129 European Parliament, 'Impacts of Shale Gas Oil Extraction on the Environment and on Human Health' (2011) at 10 http://europeecologie.eu/IMG/pdf/shale-gas-pe-464-425-final.pdf accessed 16 April 2014.

130 Poyry 'Macroeconomic Effect of European Shale Gas Production (November 2013) Where are we Now?' (2014) at 7 www.fraw.org.uk/files/extreme/poyry_2014.pdf accessed on 23 February 2014.

131 Mark Ellis-Jones (2014) Environmental Risks and Regulation of Shale Gas Exploration. The HAC Armoury London, 19 March 2014. (Notes taken by Yanal Abul Failat.)

132 LXL LLP, *A Guide to the Energy Industry for the In-House Energy lawyer: the Gas and LNG Sectors* (LXL Energy Handbook 2012) 22–23.

133 Ellis-Jones (n 131).

134 Andre Griffin, *Crisis, Issues and Reputation Management: A Handbook for PR and Communications Professionals* (Kogan Page Publishers 2014) 138.

Oil shale exploitation in the United States*

Robert A Bassett
Holland & Hart LLP

1. Introduction

Oil shale gains attention as a potential abundant source of oil whenever the price of crude oil rises. Issues of US energy independence often focus attention on oil shale, since the largest known deposits of oil shale in the world occur in the United States in the Green River Formation, which spans the states of Colorado, Wyoming and Utah. At the same time, oil shale mining and processing raise a number of environmental concerns, such as land use, availability of water, and water and air pollution. These concerns have often led to vocal public opposition.

This chapter explores the practical and legal issues associated with oil shale development in the United States.

2. What is oil shale?

Oil shale should not be confused with shale oil. 'Oil shale' is a solid rock, while 'shale oil' is a liquid. The term 'oil shale' generally refers to fine-grained, solid sedimentary rocks that contain a solid bituminous material called kerogen. Kerogen is released as petroleum-like liquids when the rock is heated in the chemical process of pyrolysis. The kerogen must be heated to more than 750 degrees Fahrenheit to convert it into oil. Once heated, the kerogen generally contains enough oil that the oil will burn without any additional processing.

Shale oil is found where the organic debris is buried deep enough through naturally occurring conditions and is exposed to temperatures sufficiently high enough to convert the kerogen into oil underground. As a liquid, shale oil can be produced through a process similar to conventional oil using horizontal drilling and fracturing.

3. Oil shale deposits in the United States

All world oil regions contain oil shale deposits; however, not all oil shale deposits represent possible economic value. Well-explored deposits, potentially classifiable as resources,[1] are found in many countries, the largest of which are in the United States.[2]

* The publishers would like to acknowledge Tasha Newland as the co-author of the version of this chapter which appeared in the previous edition

1 'Resources' refers to all oil shale deposits, while 'reserves' represents those deposits from which producers can extract oil shale economically using existing technology.

2 Other countries with oil shale deposits (listed in order of their global ranking) include Russia, the Democratic Republic of Congo, Brazil, Italy, Morocco, Jordan, Australia, Estonia, China, Canada, France, Egypt and Israel. Pierre Allix, Alan Burnham, Tom Fowler, Michael Heron, Robert Kleinberg and Bill Symington, "Coaxing Oil from Shale", *Oilfield Review* 22, no 4 (Winter 2010/2011), pp4–15.

The majority of US oil shale (and the world's largest oil shale deposit) is found in the Green River Formation in the western United States in Colorado, Wyoming and Utah[3] (see Map 1). Estimates of the oil resource in place within the Green River Formation range from 1.5 trillion to more than 2 trillion barrels.[4] Of these resources, estimates of potentially recoverable oil shale resources range from 500 billion to 1.1 trillion barrels of oil.[5]

4. Oil shale production technology

Hundreds of patents for oil shale retorting technologies exist; however, only a few dozen have undergone testing. Each technology will require varying amounts of water and energy input, and will result in different degrees of surface disturbance and resulting reclamation requirements.

To date, most oil shale exploitation from the world's oil shale has involved mining followed by shipping elsewhere, where it can be burned directly to generate electricity or processed further.[6] The extraction of the useful components of oil shale can take place above ground (*ex-situ* processing) by mining the oil shale and then treating it in processing facilities, or underground (*in-situ* processing) by applying heat and extracting the oil via oil wells.

In either *ex-situ* or *in-situ* processing, the chemical process of pyrolysis (also known as retorting or destructive distillation) is generally used to convert the kerogen in the oil shale to shale oil and shale gas. Pyrolysis is a thermochemical decomposition of organic material at elevated temperatures without the participation of oxygen. Other processes used to decompose oil shale and convert its kerogen into shale oil include hydrogenation and thermal dissolution. Hydrogenation extracts shale oil using hydrogen donors, solvents or a combination of these. Thermal dissolution involves the application of solvents at elevated temperatures and pressures.

4.1 *Ex-situ* processing

In *ex-situ* processing the oil shale is mined either underground or at the surface and then transported to a processing facility. The most common methods of surface mining involve open pit mining and strip mining. These procedures remove most of the overlying material to expose the deposits of oil shale, and become practical when the deposits occur near the surface. Underground mining generally utilises the room-and-pillar method.[7]

---

3 Ronald Johnson, US Geological Survey, "The systematic geologic mapping program and a quadrangle-by-quadrangle analysis of time-stratigraphic relations within oil shale-bearing rocks of the Piceance Basin, western Colorado", US Geological Survey Scientific Investigations Report 2012-5041 (2012).

4 James T Bartis, Tom LaTourrette, Lloyd Dixon, DJ Peterson and Gary Cecchine (2005), "Oil shale Development in the United States. Prospects and Policy Issues" (prepared for the National Energy Technology Laboratory of the US Department of Energy) at www.rand.org/pubs/monographs/2005/RAND_MG414.pdf, RAND Corporation. ISBN 978-0-8330-3848-7, (estimating 1.5 to 1.8 trillion barrels). US Department of Energy, "Oil Shale and Other Unconventional Fuels Activities", at www.fossil.energy.gov/programs/reserves/npr/npr_oil_shale_program.html (estimating more than 2 trillion barrels).

5 *Id*; see also Bureau of Land Management Colorado, "Oil Shale", at www.blm.gov/co/st/en/BLM_Programs/minerals/oil_shale.html (estimating 800 billion barrels of recoverable oil).

6 Allix *et al*, *op cit*.

MAP 1[8]

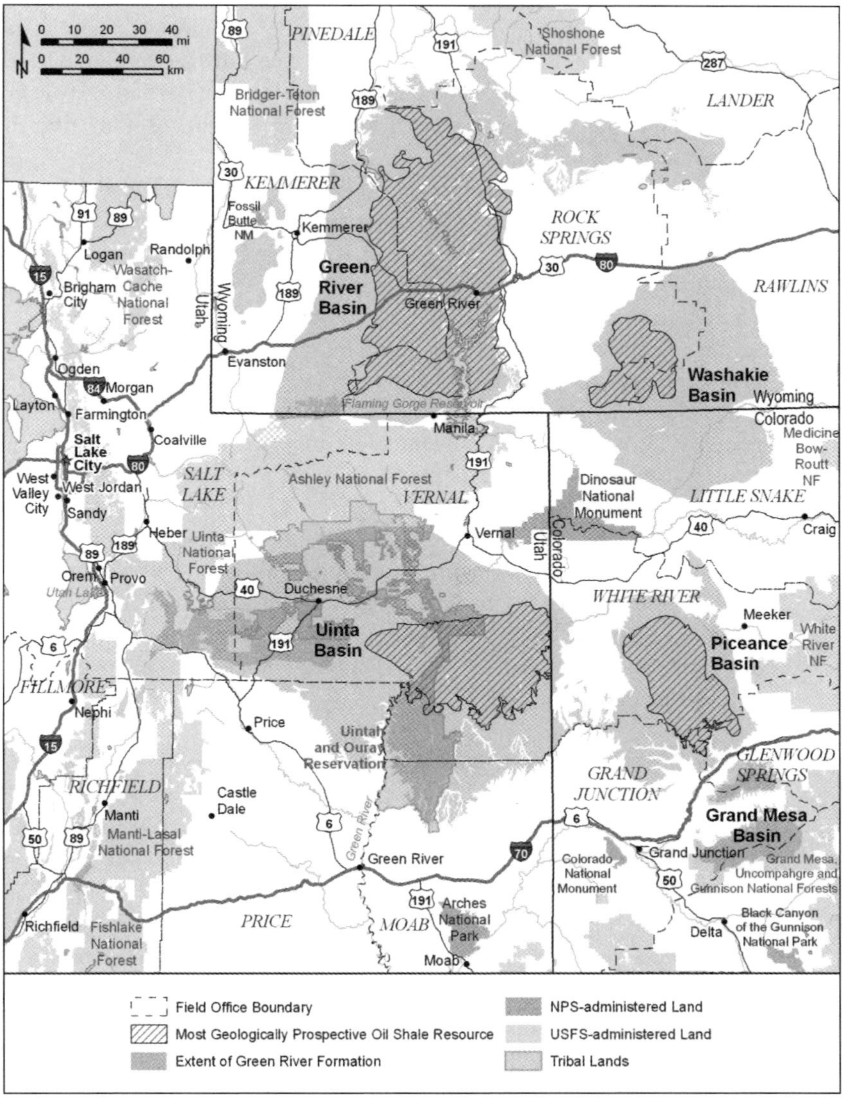

While numerous *ex-situ* processing technologies exist, two specific examples are discussed in more detail below.

7 'Room and pillar' is a mining system in which the mined material is extracted across a horizontal plane while leaving 'pillars' of untouched material to support the roof overburden leaving open areas or 'rooms' underground.

8 Source: 2012 Oil Shale & Tar Sands Programmatic Environmental Impact Statement, available at: http://ostseis.anl.gov/documents/peis2012/index.cfm.

(a) Red Leaf Resources Inc – EcoShale In-Capsule Technology

Red Leaf Resources Inc has developed the EcoShale In-Capsule Technology to produce high-quality liquid transportation fuels from oil shale, oil sands, coal, lignite and biomass. According to Red Leaf's website, the EcoShale In-Capsule Technology involves surface mining of the oil shale and then heating the mined shale in a closed surface impoundment, called a 'capsule'. The process relies on conventional mining and construction methods.[9]

According to Red Leaf, the EcoShale In-Capsule Technology protects groundwater and vegetation, uses low temperature for heating and allows for rapid site reclamation.[10] Red Leaf anticipates that approximately eight gallons (or one-fifth of a barrel) of water will be required to produce a barrel of oil using the EcoShale In-Capsule Technology. Test pilot projects reportedly support this calculation.[11] Red Leaf further anticipates that the cycle from construction to reclamation will be approximately 12 months.[12] The process also results in synthetic natural gas production that will provide for energy self-sufficiency for ongoing heating of materials in the capsule after the initial heating has been completed. This process is estimated to produce a commercially attractive return on investment at oil prices greater than about $45 a barrel of oil.[13]

Red Leaf has performed simulation and mathematical modelling of its technology and reportedly confirmed results with bench testing and external lab analysis of its produced hydrocarbon products.[14] Red Leaf also completed a field pilot test of the EcoShale In-Capsule Technology in the Uintah Basin, Utah. According to Red Leaf, the test validated the technology modelling and most commercial engineering design aspects.[15]

(b) Colony Shale Oil Project – TOSCO II retorting technology

The Colony Shale Oil Project was an oil shale development project established in 1963 in the Piceance Basin in Colorado.[16] Over time, various companies acquired and sold an interest in the project. By 1970 ownership was vested 60% in Exxon Corporation and 40% in TOSCO Corporation. In the 1960s a prototype mine and plant operation proved the feasibility of the underground mining plan with above-ground processing using the TOSCO II retorting technology, developed by TOSCO.[17] The TOSCO II process would use fine particles of oil shale that are heated in a rotating kiln or drum. In this hot recycled solids process, hot ceramic balls would be

---

9 Red Leaf Resources Inc, "EcoShale In-Capsule Technology", at www.redleafinc.com.
10 *Id.*
11 Red Leaf Resources Inc, "Oil Shale FAQs", at www.redleafinc.com.
12 *Id.*
13 *Id.*
14 US Department of Energy, Office of Petroleum Reserves, Naval Petroleum and Oil Shale Reserves, *Secure Fuels from Domestic Resources. Profiles of Companies Engaged in Domestic Oil Shale and Tar Sands Resource and Technology Development* (5th ed) (September 2011).
15 *Id.*
16 BLM, Draft Programmatic Environmental Impact Statement and Possible Land Use Plan Amendments for Allocation of Oil Shale and Tar Sands Resources on Lands Administered by the Bureau of Land Management in Colorado, Utah and Wyoming, Appendix A-1 (January 2012), available at http://ostseis.anl.gov/documents/peis2012/index.cfm.
17 *Id.*

used for the heat transfer.[18] The design for the commercial Colony Shale Oil Project operations included an underground room-and-pillar mine with above-ground retorting and upgrading facilities.[19] Approximately 200,000 barrels[20] of shale oil were produced experimentally, but no shale oil was ever produced commercially.

Exxon terminated the project on 2 May 1982, a day known among locals as 'Black Sunday' because of low oil prices and increased expenses resulting in more than 2,000 workers being laid off.[21]

4.2 *In-situ* processing

In-situ processing converts the kerogen to a liquid while it is still underground in the form of an oil shale deposit, following which the liquid is extracted through wells similar to conventional crude oil wells. Such technologies can potentially extract more oil from a given area of land than *ex-situ* processes, since they can access the material at greater depths than surface mines. True *in-situ* processes do not involve mining or crushing the oil shale. Modified *in-situ* processes involve drilling and fracturing the target oil shale deposit to create voids in the deposit to enable a better flow of gases and fluids through the deposit.[22]

While numerous *in-situ* processing technologies exist, two specific examples and one theoretical process are discussed in more detail below.

(a) *Shell* – In-situ *Conversion Process*

Between 2003 and 2005 Shell carried out a small field test known as the Mahogany Demonstration Project South on its private property in Rio Blanco County, Colorado using an in-ground heating process to recover oil and gas from the shale formation, referred to as the In-situ Conversion Process.[23]

The In-situ Conversion Process requires a freeze wall to be constructed to isolate the processing area from surrounding groundwater. The freeze wall is constructed by drilling 2,000-feet deep wells approximately eight feet apart and utilising a closed-loop pipe system with circulating super-chilled ammonia liquid to cool the ground.[24] This system causes the water in the surrounding rock to freeze and form an underground wall of ice. The freeze wall serves as an impermeable barrier to keep groundwater out of the production zone.[25]

After installation of the freeze wall, the production zone is dewatered and heating and recovery wells are drilled at 40-foot intervals within the working zone. Electrical heating elements are lowered into the heating wells and are used to heat

18 US Department of Energy, Office of Petroleum Reserves – Strategic Unconventional Fuels, "Fact Sheet: Oil Shale Conversion Technology", available at www.unconventionalfuels.org/factsheets/Oil_Shale_Technology_Fact_Sheet.pdf.
19 *Id.* Bartis *et al*, *op cit.*
20 One barrel of shale oil is equal to 42 gallons.
21 Bartis *et al*, *op cit.*
22 US Department of Energy, Office of Petroleum Reserves – Strategic Unconventional Fuels, "Fact Sheet: Oil Shale Conversion Technology", available at www.unconventionalfuels.org/factsheets/Oil_Shale_Technology_Fact_Sheet.pdf.
23 Shell, United States, "Technology", at www.shell.us.
24 *Id.*
25 *Id.*

the oil shale over a period of two years or more.[26] Through the heating process, the kerogen in the oil shale is slowly converted into shale oil and gases, which then flow to the surface through the recovery wells. The residual shale that remains underground following production of shale oil and gases is then cleaned of contaminants by flushing with clean water.

Shell has conducted at least six oil shale heating tests and two freeze wall tests on its private property.[27] To date, Shell's research has demonstrated that its In-situ Conversion Process works technically on a small scale, but it still needs to prove that it can work commercially.[28] Shell also conducted a successful freeze wall test which proved its ability to isolate an area to be heated, protect surrounding groundwater and constantly monitor and verify the freeze wall integrity.[29]

Shell's goal is to advance the In-situ Conversion Process technology systematically to the point at which an application could be made to convert its 160-acre research, development and demonstration tracts to commercial leases.[30] Shell anticipates making a commercial decision in the middle of the next decade or possibly later, depending on the sequence and outcome of research activities.[31]

Shell's In-situ Conversion Process will require high volumes of water, mainly to refine the oil and purify the natural gas, but also for the associated underground and surface uses described below. Certain parties opposed to Shell's project have asserted that the *in-situ* nature of Shell's process might pose a risk of contaminating underground aquifers. The anticipated high volumes of water required and the risk of groundwater pollution are the technology's greatest challenges.[32]

(b) American Shale Oil LLC – Conduction, Convection, Reflux process

American Shale Oil LLC (AMSO) proposes to use its 'Conduction, Convection, Reflux' process to recover shale oil with minimal surface disturbance, minimal water usage and minimal imported energy.[33] AMSO's concept will use a combination of parallel heating and production wells to create a panel of retorts covering roughly 2,000 square feet.[34] The panels will be separated by a relatively thin strip of unheated shale that would create a permeability barrier between the panels. The heat for converting kerogen to oil and gas will be supplied by downhole burners in long horizontal wells near the base of the oil shale bed to be retorted.[35]

The Conduction, Convection, Reflux process will be separated from the aquifers by multiple geologic barriers, including 100 to 300 feet of impermeable nahcolite-

---

26 BLM, Draft Programmatic Environmental Impact Statement and Possible Land Use Plan Amendments, *op cit.*

27 Peter M Crawford, Christopher Dean and Jeffrey Stone, INTEK, Inc and James C Killen, US Department of Energy, "Assessment of Plans and Progress on US Bureau of Land Management Oil Shale RD&D Leases in the United States" (29 April 2012).

28 Shell, United States, "Technology", at www.shell.us.

29 US Department of Energy, Office of Petroleum Reserves, Naval Petroleum and Oil Shale Reserves, *op cit.*

30 Shell, United States, "Technology", at www.shell.us.

31 *Id.*

32 Jon Birger, "Oil shale may finally have its moment", *CNN Money*, available at http://money.cnn.com/2007/10/30/magazines/fortune/Oil_from_stone.fortune/index.htm.

33 American Shale Oil LLC, "Our Concept", at www.amso.net/Our-Concept.aspx.

34 *Id.*

35 US Department of Energy, Office of Petroleum Reserves, Naval Petroleum and Oil Shale Reserves, *op cit.*

rich oil shale to avoid aquifer contamination.[36] In addition, AMSO will establish a dewatered zone in the retorting zone by installing pumping wells surrounding the subsurface retort area and re-injecting extracted groundwater down-gradient into the equivalent aquifer intervals in order to maintain the regional water table and avoid disturbing baseflow to nearby streams.[37] After project completion, pumping and treating of contaminated groundwater may be necessary. AMSO's anticipated water usage is less than one barrel of water per barrel of shale oil produced.[38]

AMSO holds a research, development and demonstration lease from the US government for 160 acres in western Colorado. Site characterisation activities by AMSO demonstrated hydrologic isolation of the oil shale deposit from protected underground water sources.[39] Drilling began on AMSO's pilot test production, heating and monitoring wells in March 2011. The pilot test will include developing and demonstrating methods to enhance environmental protections.

(c) *Nuclear explosions*

In the 1960s and early 1970s the US Atomic Energy Commission, a predecessor agency of the US Department of Energy, conducted a series of experiments under the Plowshare Programme, which was designed to develop peaceful uses for nuclear energy.[40] One such experiment was Project Rulison in 1969, which involved detonation of a 40-kiloton nuclear device 8,426 feet below the ground surface in an attempt to release commercially marketable quantities of natural gas from the fine-grained, low-permeability sandstone of the Williams Fork Formation in western Colorado, near the Green River Formation.[41]

The objective of Project Rulison was to stimulate the flow of natural gas through fractures created by the blast and use the chimney as a collection chamber.[42] While natural gas was produced through a re-entry well drilled into the top of the chimney, elevated levels of radioactivity in the gas made it unacceptable for use at that time.[43] Contamination at the Rulison site includes radioactive contamination of the deep bedrock around the shot cavity, possible surface contamination from the gas flaring and decontamination activities, and near-surface hazardous waste contamination from the closed mud pits.[44] The primary contaminant of concern remaining in the deep subsurface is tritium, a mobile radioisotope of hydrogen.[45]

The Department of Energy reports that no radioactive contamination associated with the Rulison test has been detected in any samples taken from the nearby municipal drinking water supply springs, water supply wells, springs or gas and water

---

36 *Id.*
37 BLM, Draft Programmatic Environmental Impact Statement and Possible Land Use Plan Amendments, *op cit.*
38 American Shale Oil LLC, "Our Concept", at www.amso.net/Our-Cencept.aspx.
39 AMSO, "Oil Shale and the AMSO RD&D Program" (October 2011), available at www.amso.net/amso/files/f4/f49448bc-7ce1-4ab7-b637-f8890fc75878.pdf.
40 US Department of Energy, Legacy Management, "Final Rulison Path Forward" (June 2010).
41 US Department of Energy, Legacy Management, "Rulison, Colorado, Site. Fact Sheet" (5 October 2011).
42 *Id.*
43 *Id.*
44 US Department of Energy, Environmental Management, Rio Blanco and Rulison Sites, at www.em.doe.gov/bemr/BRMRSites/prbs.aspx.
45 *Id.*

vapour from nearby gas production wells.[46] Nonetheless, the Environmental Protection Agency and the Department of Energy continue to monitor and manage the site to ensure protection of human health.[47] Furthermore, federal and state institutional controls impose certain restrictions on drilling near the Rulison site.[48]

In the 1960s a proposal was suggested for a modified *in-situ* oil shale extraction process which involved a process very similar to the Rulison experiment.[49] In essence, the proposal included:

- creation of a rubble nuclear chimney in the oil shale through a deep underground nuclear explosion;
- *in-situ* retorting in the chimney fuelled by the energy from the nuclear detonation itself and the subsequent controlled combustion of residual carbon in retorted oil shale; and
- production through conventional production wells.[50]

This approach was abandoned for a number of technical reasons, but could be resurrected through technological advancements.

One company that hopes to resurrect the use of nuclear power in oil shale development is Shale and Sands Oil Recovery LLC (SASOR). SASOR holds a patent on the use of nuclear energy technology for the extraction of oil from oil shale and anticipates collection of licensing and royalty fees from its use in the future.[51]

5. Legal issues related to oil shale exploitation

Commercial oil shale production requires overcoming numerous practical considerations which have serious related legal implications. Such issues include acquisition of mineral rights tenure, water rights and water usage requirements, and environmental permitting.

5.1 Mineral rights tenure[52]

The first step in exploiting oil shale is gaining access to the resource. Oil shale minerals in the United States can be owned by the federal government, the state government, private parties or Native American tribes. The ownership of the minerals will dictate the possible development scenarios and associated legal requirements. However, more than 70% of the Green River Formation oil shale resources lie beneath federal lands, primarily lands managed by the US Bureau of Land Management (BLM).[53] The other 30% lie beneath state and private lands.

---

46 *Id.*

47 *Id.*

48 *Id*; US Department of Energy, Legacy Management, "Final Rulison Path Forward" (June 2010).

49 DB Lombard, "Recovering Oil by Retorting a Nuclear Chimney in Oil Shale", *Journal of Petroleum Technology* (June 1967).

50 *Id.*

51 US Department of Energy, Office of Petroleum Reserves, Naval Petroleum and Oil Shale Reserves, *op cit.*

52 This section relies heavily on writing by Jean Feriancek and Karol Kahalley of Holland & Hart LLP, whom the authors thank for their contribution.

53 US Department of the Interior, Bureau of Land Management, "Oil Shale Resources on Public Lands", at www.blm.gov/wo/st/en/prog/energy/oilshale_2/background.html (updated 12 April 2011).

(a) *Federal government ownership of lands and minerals*

The federal government is the largest landowner in the area covered by the Green River Formation. Under federal law, oil shale was originally considered a 'hard rock' mineral that could be obtained by locating a mining claim (similar to claims for gold and precious metals). Under the General Mining Law of 1872,[54] a US citizen or company can obtain the right to prospect for, mine and remove valuable mineral deposits by 'locating' a mining claim on open public domain land (ie, land owned by the federal government and not closed to mineral location by executive order or otherwise) and thereafter maintaining that mining claim in effect in accordance with the statute. If a patent is issued for a mining claim, title to the land on which the claim was located passes from the United States to the party to which the patent (a form of deed) was issued. Patenting is not required, however, and under the General Mining Law the holder of a valid unpatented mining claim has the right to mine and remove minerals without ever patenting the claim.

Originally it was possible to locate mining claims under the General Mining Law for lands containing oil shale deposits. In 1920 Congress enacted the Mineral Leasing Act of 1920, which placed oil shale under a leasing system along with all of the fuel minerals (eg, oil, gas and coal).[55] Thereafter, such lands were no longer open to location and acquisition of title under the General Mining Law, but only to lease. As the US Supreme Court stated in *Hickel v Oil Shale Corp* (known as the 'TOSCO' case), enactment of the Mineral Leasing Act "completely changed the national policy over the disposition of oil shale lands".[56] However, the act contained a savings clause that protected all "valid claims existent" on 25 February 1920, which were "thereafter maintained in compliance with the [General Mining Law] under which initiated, which claims may be perfected under such laws, including discovery".[57] Thus, while no new oil shale mining claims could be located after 25 February 1920, and oil shale was no longer subject to private appropriation after that date, numerous unpatented oil shale mining claims that had been located on public domain land before that date and not previously patented or abandoned remained in effect after that date.

In 1976 Congress enacted the Federal Land Policy and Management Act of 1976,[58] which modified the existing laws related to public lands. The act established a federal filing system for unpatented mining claims and imposed filing requirements on the owner of an unpatented mining claim.[59] It provided that any claims that did not meet the filing requirement were void *ab initio*, and the owners of many historic mining claims for oil shale saw their claims declared void for failure to make the required filings.

The Energy Policy Act of 1992[60] placed significant limitations on patents of oil shale mining claims. Only a holder of a valid oil shale claim which had filed a patent application before 24 October 1992 and proceeded to a certain stage in the patenting

---

54 30 USC Section 22 and following.
55 30 USC Section 181 and following.
56 See *Hickel v Oil Shale Corp*, 400 US 48, 51 (1970).
57 30 USC Section 193.
58 43 USC Section 1701 and following.
59 43 USC Section 1744.
60 30 USC Section 242.

process (ie, received a 'first half final certificate') could continue with the patenting process under the General Mining Law. It created a category of 'limited patent', which grants title to oil shale and associated minerals only and does not include title to the surface or other minerals.

Federal investigation/leasing programme: In September 2008 the BLM issued a Proposed Plan Amendments/Final Oil Shale and Tar Sands Programmatic Environmental Impact Statement (2008 PEIS) pursuant to Section 369 of the Energy Policy Act, the Federal Land Policy and Management Act, and the National Environmental Policy Act of 1969. The 2008 PEIS analysed the environmental and socioeconomic impacts of amending 12 land use plans in Colorado, Utah and Wyoming to designate public lands administered by the BLM as available for commercial leasing for oil shale or tar sands development. The record of decision that followed adopted the proposed land use amendments, reflecting the allocation decisions analysed in the 2008 PEIS. In doing so, the BLM made approximately 2 million acres of BLM-managed lands within the Green River Formation available for potential development of oil shale.

These land allocation decisions were challenged in a lawsuit brought by a coalition of environmental organisations in January 2009. As part of a settlement agreement entered into by the United States to resolve the lawsuit, the BLM re-analysed the land allocations in the 2008 PEIS to consider excluding certain lands from future leasing of oil shale and tar sands resources. Specifically, the BLM considered amending the applicable resource management plans to specify whether any areas in Colorado, Utah and Wyoming should be made available for leasing and development of oil shale or tar sands.

The BLM issued a record of decision in 2013 (2013 ROD)[61] that amended 10 land use plans in Colorado, Utah and Wyoming to make approximately 678,000 acres available for potential development of oil shale, and approximately 132,000 acres available for development of tar sands. The 2013 ROD included public lands administered by the BLM where the federal government owns both the surface estate and sub-surface mineral rights, and where the federal government owns the sub-surface mineral rights but the surface estate is owned by tribes, states or private parties. The 2013 ROD also contained the BLM's analysis of the direct, indirect and cumulative environmental, cultural and socioeconomic impacts of the revisions to its land allocation decision.

The 2013 ROD provided that the areas allocated as open for future oil shale leasing are open only to research, development and demonstration (RD&D) leases. The BLM will issue a commercial lease only when a lessee satisfies the conditions of its RD&D lease and the regulations in the Code of Federal Regulations, Title 43, Subpart 3926 for conversion to a commercial lease. The preference right acreage, if any, which would be included in the converted lease would be specified in the RD&D lease.

---

61 BLM, Approved Land Use Plan Amendments/Record of Decision (ROD) for Allocation of Oil Shale and Tar Sands Resources on Lands Administered by the Bureau of Land Management in Colorado, Utah and Wyoming and Final Programmatic Environmental Impact Statement, March 2013.

The land use plan amendments remove from potential oil shale and tar sands leasing several categories of land within the planning area in Colorado, Utah and Wyoming. In total, more than 1,340,770 acres of the planning area in Colorado, Utah and Wyoming are excluded from oil shale leasing and development, and more than 301,100 acres in Utah are excluded from tar sands leasing and development.

If and when applications to lease are received and accepted for oil shale or tar sands resources within the acres available for leasing under the 2013 ROD, the BLM will conduct additional required analyses, including consideration of direct, indirect and cumulative effects of the proposed development, reasonable alternatives, and possible mitigation measures. On the basis of that analysis of future lease application(s), the BLM will establish general lease stipulations and best management practices, and amend applicable land use plans, if necessary. After a lease is authorised, actual development will require additional analysis to address the site-specific conditions of the proposed development and to develop mitigation measures as necessary.

(b) State government ownership of land and minerals

Many state lands and mineral interests are within the area covered by the Green River Formation. Individual states have obtained ownership of their lands and the underlying minerals by a variety of means, including:

- explicit grants from the federal government;
- direct succession of the ownership of a foreign nation (eg, the original 13 states and Texas); and
- implication as to their sovereignty as a state (eg, lands and minerals beneath inland navigable waters and certain offshore lands).

The laws of each state vary as to whether state-owned mineral rights may be disposed of by sale, lease, location or other means. State lands and minerals are generally available for leasing for mineral development from state governments. State-owned lands and minerals are managed in Colorado by the Colorado State Land Board, in Wyoming by the Office of State Lands and Investments, and in Utah by the Utah School and Institutional Trust Lands Administration.

(c) Private ownership of land and minerals

Land and the underlying minerals in the United States may be owned in fee by individuals or other legal entities. Furthermore, surface land and minerals under the surface may be owned as separate property estates in what is known as a 'split estate'. There are many split-estate lands in the area of the Green River Formation, as well as lands where the surface owner owns all of the underlying minerals. The ownership of surface land and minerals may be freely conveyed, transferred, mortgaged and encumbered by private citizens and entities by any legal means. Because the surface estate and the mineral estate may be owned as separate and distinct property estates, the surface and minerals may be treated as separate estates for the purposes of such conveyances.

While the mineral estate was historically considered the dominant estate, the accommodation doctrine has been adopted in many states. This doctrine provides that the mineral rights owner has an implied easement to use as much of the surface as is reasonable and necessary to develop the minerals, and that the surface owner and the mineral owner must each accommodate and not interfere with or damage the other owner's rights to use the surface. The exact extent of what constitutes 'reasonable and necessary use' is generally defined by state case law; typically, a mineral developer must reach agreement with the surface owner regarding surface use before undertaking any development of minerals where the mineral owner does not own or otherwise control the surface.

(d) *Native American ownership of lands and minerals*

There are many lands in the area of the Green River Formation owned by Native American tribes or individual Native Americans, particularly in Utah. Lands owned by Native American tribes are held generally for the beneficial ownership of all members of the tribe, with individual tribal members having no individual ownership interest in such lands. Tribal land ownership generally includes ownership of all minerals in those lands. In limited instances, the lands and minerals are also owned by individual Native Americans under antiquated allotment statutes. Tribes and individual Native American allottees may enter into agreement with private parties for the lease of their minerals for exploration, development and production, and for surface access to such resources. Such contracts are subject to approval by the US secretary of interior, pursuant to the federal government's fiduciary relationship with the tribes. Tribal law typically governs all contracts with tribes and allottees.

5.2 Water rights and water usage requirements

The availability of water may be the biggest legal and technical constraint on oil shale development on lands in the area of the Green River Formation. All high-grade western oil shale resources in the Green River Formation lie in the Colorado River's drainage basin (which covers the majority of lands in the area of the formation). Most current oil shale technology needs large amounts of water for production and processing, depending on the technology employed. Water may be needed for:

- dust control during materials extraction, crushing and transport;
- cooling and reclaiming spent shale;
- raw shale oil upgrading;
- various plant utilities associated with power production and environmental control;
- hydraulic fracturing and oil and natural gas extraction; and
- groundwater remediation.

According to the RAND Corporation, *ex-situ* retorting processes are estimated to use between two and five barrels of water per barrel of produced shale oil.[62] *In-situ* retorting eliminates or reduces a number of these water requirements, but considerable volumes of water may still be required.

---

62 Bartis *et al, op cit.*

The Colorado Water Conservation Board – the Colorado state agency overseeing Colorado's water resources – prepared a report in 2011 analysing the state's water needs until 2050 and estimated that an oil shale industry producing 1.56 million barrels of oil a day could use between 0 and 120,000 acre-feet of water a year, depending on what technologies and other factors are implemented.[63] The report also projected that by 2050, water use ranges would be from 0 to 44,000 acre-feet of water a year for an industry providing 550,000 barrels of oil a day.[64]

The Colorado River drainage basin's water resources are already tightly regulated and in great demand. Rapid population growth, rising demand for electric power, growing recreational use and increased efforts to maintain and restore the river's ecosystems all contribute to the demands placed on the basin. Major oil companies with interest in oil shale development (eg, Total E&P, Exxon Mobil Corp, Royal Dutch Shell PLC and Chevron Corp) have already started purchasing senior water rights from mostly agricultural users in the Piceance Basin, which lies between the White and Colorado Rivers within the Colorado River drainage basin.[65]

In Colorado, a separate Water Court system has jurisdiction to approve, via decree, legal water rights for new uses of water and changes of use for existing water rights. The process of identifying and securing available water rights, confirming the physical availability of water in sufficient quantities and perfecting the water rights in the Water Court for oil shale operations and power generation can be time consuming and resource intensive, particularly if significant opposition from other water rights holders exists.

Rights to the use of water in the state of Utah must be established through the appropriation process administered by the Division of Water Rights. Many areas of Utah are administratively 'closed' to new appropriations of water. In those areas, new diversions and uses of water are established by the modification of existing water rights. Such modifications are accomplished by the filing of change applications. In Wyoming, a permit system administered by the Wyoming State Engineer's Office governs the allocation, diversion and use of water rights.

5.3 Environmental permitting

Current and future environmental regulations present serious risks to developers of oil shale deposits. Oil shale mining and processing activities will require various federal, state and local permits that govern different aspects of the process, ranging from mining permits, water quality permits, air quality permits, county permits and permits related to wildlife protection statutes, to permits required by county and local governments. The implications of climate change regulations may also pose

63 Colorado Water Conservation Board (26 January 2011), Statewide Water Supply Initiative 2010 Mission Statement, Key Findings, and Recommendations, available at http://cwcb.state.co.us/water-management/water-supply-planning/Documents/SWSI2010/SWSI2010FactSheet.pdf. Retrieved 10 July 2012.

64 *Id.*

65 Scott Streater, "Population growth, oil shale activity portend huge deficits for Colo", *Land Letter: The Natural Resources Weekly Report* (13 January 2011), available at www.eenews.net. Western Resource Advocates, "Water on the Rocks. Oil shale Water Rights in Colorado" (2009), available at www.westernresourceadvocates.org/land/wotrreport/index.php.

additional air quality permitting hurdles to an oil shale mining or processing operation, since the production of petroleum products derived from oil shale may entail significantly higher emissions of carbon dioxide compared with conventional crude oil production and refining. Disposal of spent shale will require specific environmental permits, depending on the disposal method. In addition, where applicable, the National Environmental Policy Act permitting process – which requires preparation of an environmental impact statement or environmental assessment to analyse the potential direct, indirect and cumulative impacts of a proposed federal action, including leasing by the BLM, and reasonable alternatives to the proposed action – may be triggered. The specific permitting requirements for a proposed oil shale development project will depend on many factors, including the technology to be implemented and the location of the activities.

A public notice and comment period is required in order to obtain many of the necessary permits for an oil shale mining or processing facility. Exhibit 1 shows the Colorado Oil Shale Permitting Process, Timeline and Fact Sheet, published by the Colorado Division of Reclamation Mining and Safety, which identify the requirements to obtain a permit to mine for oil shale in Colorado.

The public notice and comment period lengthens the time required to obtain such permits and gives public and environmental organisations an opportunity to submit comments in opposition to the proposed project. Due to the controversial nature of oil shale development in the United States, it is likely that environmental organisations and other interested parties such as farmers and ranchers will bring litigation challenging the issuance of any permit related to oil shale production. Critics of oil shale development have been active in Utah and Colorado.

EXHIBIT 1[66]

Timeline	Colorado oil shale permitting process – timeline and fact sheet
Pre-filing	**Delivery of Application:** The first step in the Division of Mining and Reclamation (DRMS) permitting process for the Research, Development and Demonstration (RD&D) site requests is the delivery of the application. At that time, the applicant also is required to post a sign at the designated site, and deliver a copy to the local county clerk and recorder. **Initial Administrative Review:** DRMS performs an initial review of the application, and must notify the applicant of any deficiencies within 10 working days. Applicant has 60 days to rectify any deficiencies and submit needed documents, or face application denial.

continued on next page

66 From www.mining.state.co.us/DNR_Process%20Flow%20Chart_Revised.pdf.

Timeline	Colorado oil shale permitting process – Timeline & fact sheet
Filing date	**Application Considered Filed:** Once the application is filed and considered ready for technical review notices are sent to the applicant and all necessary federal, state and local agencies. Public Comment Period begins, and comments must be submitted to DRMS in writing.
Days +/-10 4 to 48	**Publication of Public Notice:** Within 10 days of filing, applicant begins weekly publication of public notice for four consecutive weeks in a local newspaper. Immediately following first publication, the applicant must mail a copy of the notice to all landowners adjacent to or within 200 feet of the site.
Days 48 to 58	**Public Comment Period:** The public comment period ends 20 days after final published public notice. Objection letters received by DRMS are sent to the applicant within 10 days of receipt.
Days 49-150	**Notices of Pre-hearing Conference and Formal Board Hearing:** These meetings are only held if objections are received during the public comment period. DRMS issues notice of these meetings to all parties. A 30-day notice is provided in advance of the Formal Board Hearing. **DRMS Recommendation for Approval or Denial:** This recommendation is due 150 days after the filing date unless the applicant extends the date. An extension can be granted for up to 365 days from the filing date. If there are no objections, the DRMS decision ends the application process.
Days 47-167	**DRMS Recommendation Rationale Notice:** DRMS must mail the recommendation and rationale for approval or denial to all parties at least three working days prior to the Pre-hearing Conference.
Days 50-170	**Pre-hearing Conference:** This conference clarifies the issues to be discussed at the Formal Board Hearing and helps set the hearing agenda. Participating parties must provide witness lists and exhibits at this conference. This conference is held a minimum of 10 days prior to the Formal Board Hearing.

continued on next page

Timeline	Colorado oil shale permitting process – Timeline & fact sheet
Days 52-172	**Motions Filing Period:** Motions must be filed by participating parties within two working days following the Pre-hearing Conference.
Days 78-179	**Pre-hearing Conference Order:** The Pre-hearing Conference order and hearing agenda is made public to all parties prior to Formal Board Hearing.
Days 76-177	**Motions Response:** The Board must receive responses to all motions at least three working days prior to Formal Board Hearing.
Days 79-180	**Formal Board Hearing:** Formal Board Hearing on the application is held at least 10 days following the Pre-hearing Conference, but before the 180th day after the filing date. At this meeting, Board members will render a decision on the application.

6. Conclusion

The United States, like many other jurisdictions, contains a vast amount of oil shale deposits. Attempts at oil shale development have a long history, but the commercial production of oil shale deposits is still in its infancy. Navigating the legal issues associated with development of an oil shale project in the United States will present many challenges. The advancement of oil shale production technology will lead to commercial development opportunities and evolving regulatory requirements. Acquiring the necessary mineral tenure, water rights and permits will be costly and time consuming. Additional regulations, with heightened environmental awareness, local opposition and market price fluctuations, will present businesses with many legal obstacles to project development.

Unconventional gas in Australia

Sharon Eacott (Wilson)
Alinta Energy
Belinda Fan
Herbert Smith Freehills

1. Introduction

An expansion in worldwide demand for energy and an ever-growing awareness of the environmental impact of greenhouse gases has seen countries around the world turn to cleaner and renewable sources of energy. Australia, with its abundant reserves of natural gas, including a growing industry in coal seam gas and shale gas, is at the forefront of this movement.

Much attention has been paid to the booming coal seam gas industry on the east coast of Australia, fuelled by the commencement of three coal seam gas liquefied natural gas projects in Queensland.

These Queensland projects have fundamentally changed the east coast's gas industry. Their parallel commencement in the last two years has created upward pressure on domestic gas prices. Rising gas prices and liquid natural gas export potentials led to a short spike in interest in Australia's 'other' unconventional gas resources – shale gas and tight gas; however, this interest has recently been checked by the impact of the oil price slump.

The transition of coal seam gas to a conventional source of energy in Australia, like shale gas in the United States, has not been without its challenges. The exploitation of coal seam gas and shale gas has received much public attention and is an area that has been subject to ongoing debate and potential legislative reform. New South Wales, Victoria and the Northern Territory have recently introduced or announced the introduction of measures designed to curb unconventional gas exploration and development.

This chapter provides an overview of the unconventional gas landscape in Australia with a discussion of the multifaceted legal, commercial, social and environmental issues facing the industry.

2. What is coal seam gas?

This chapter covers both types of unconventional gas found in Australia: coal seam gas[1] (CSG) and tight sands gas/shale gas. Given that other chapters in this book have already covered the characteristics of shale gas, this section will primarily focus on CSG. A comparison of the characteristics of CSG and shale gas will provide background to a discussion of the issues specific to CSG projects.

1 Also known as coal bed methane.

CSG is a form of naturally occurring gas, consisting primarily of methane (generally 95%–97% pure methane) that is trapped in underground coal seams by the pressure of formation water in the coal cleats and fractures.[2] CSG, unlike shale gas, is found meshed with the matrix of coal in the underground seams and held in place by underground water pressure. CSG is found closer to the surface than shale gas, typically at depths of 300–1,000 metres below the surface.[3] Shale gas on the other hand is found at significantly greater depths, usually beyond 1,500 metres.[4]

CSG is extracted through wells which are drilled into coal seams. To release gas from the coal, formation water is removed from the coal seams to the surface.[5] The flow of CSG to the surface and the total amount of gas extracted is often increased by a stimulation process called hydraulic fracturing or 'fracking'.[6] Hydraulic stimulation was first commercially established in the late 1940s in the United States[7] and has been used to enhance the production of CSG in Australia since the 1990s.[8] The fracking process involves encasing the well from top to bottom with steel casing. To gain access to the coal, the casing is perforated at specific intervals along the well. Fluid which is typically composed of water, sand and chemical additives is injected at high pressure into the cased well to create fractures on the coal seam where the well is perforated.[9] After the injection stops, sand and additives remain in place to hold the fractures open and to form a conductive channel in the coal through which water and gas can flow back to the well.[10]

Both the water and the CSG will flow through the drilled well to the surface, where it will be separated. The gas will be sent to be stored under compression and the water will be treated to ensure it meets water quality standards before it is injected into underground water systems for other purposes such as irrigation, watering livestock or commercial use.[11]

Unlike shale gas, CSG production requires 'produced water' to be pumped from the well throughout the entirety of the operating life of the project. This means that there is a significantly larger quantity of produced water that will need to be dealt with by CSG producers than in a shale gas project.

---

2 "What is unconventional gas?", CSIRO, (25 February 2015) www.csiro.au/en/Research/Energy/Hydraulic-fracturing/What-is-unconventional-gas viewed 22 September 2016.

3 *Ibid.*

4 *Ibid.*

5 "Coal seam gas water", Department of Environment and Heritage Protection QLD, https://www.ehp.qld.gov.au/management/non-mining/csg-water.html.

6 "What is unconventional gas?", CSIRO, see footnote 4.

7 "Implications for Western Australia of Hydraulic Fracturing for Unconventional Gas", Standing Committee on Environment and Public Affairs, (7 November 2015), 23 www.parliament.wa.gov.au/Parliament/commit.nsf/(Report+Lookup+by+Com+ID)/74E61E739E39E57748257EF9002150FE/$file/ev.fra.151117.rpf.042.xx.pdf.

8 "What is hydraulic fracturing?", CSIRO, (24 February 2015) www.csiro.au/en/Research/Energy/Hydraulic-fracturing/a-What-is-hydraulic-fracturing.

9 *Ibid.*

10 *Ibid.*

11 "General beneficial use approval – Associated water (including coal seam gas water)", Department of Environment and Heritage Protection QLD, (2014) https://www.ehp.qld.gov.au/assets/documents/regulation/wr-ga-associated-water.pdf.

3. Unconventional gas in Australia

3.1 Australia's unconventional gas reserves

In 2015, Australia accounted for approximately 1.9% of world gas reserves and 1.9% of world gas production.[12] Australia's total identified gas resources are reported to be around 279,819 petajoules (PJ); an amount that has been touted to be equal to approximately 106 years of gas at current production rates, of which the gas reserves account for 47 years.[13] Within this figure is a significant level of unconventional gas resources including CSG, tight gas and shale gas (CSG accounts for 79,583 PJ).[14]

The map below describes Australia's conventional and CSG petroleum-producing basins.

Location of Australia's gas resources and infrastructure

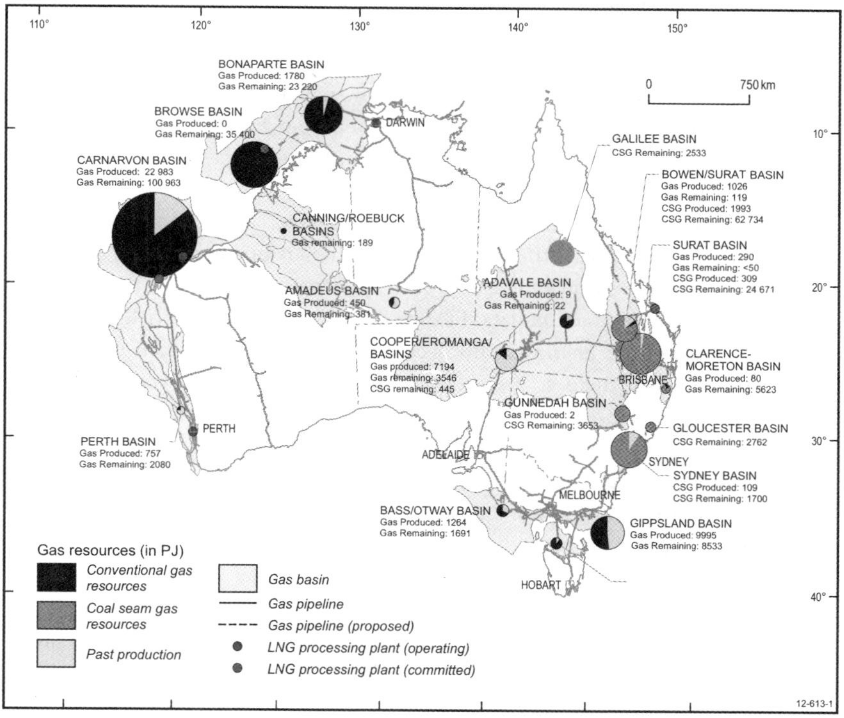

Source: Geoscience Australia, Australian Gas Resource Assessment 2015

Most of Australia's significant conventional gas resources are located in the offshore areas in north western Australia (Carnarvon, Browse and Bonaparte basins) and in south eastern Australia (Otway, Bass and Gippsland basins), and in onshore

---

12 "Statistical Review of World Energy – June 2016", BP, (June 2016) www.bp.com/statisticalreview.

13 "Australian Energy Resources Assessment: Gas – Summary" Geoscience Australia, (2014), www.ga.gov.au/aera/gas.

14 *Ibid.*

central Australia (Cooper, Eromanga and Warburton basins).[15] The major CSG resources are based in the coal basins of Queensland, New South Wales and South Australia.[16] Shale gas accumulations are reportedly located in Queensland, Northern Territory, South Australia and Western Australia.[17]

The development of new technology in horizontal drilling and hydraulic fracturing, and the increasing demand from Australian domestic markets has seen unconventional gas feature as a significant contributor to the commercial viability of Australia's gas resources.[18] According to recent industry reports, unconventional gas accounts for 88% of 2P east coast gas reserves and 90% of 3P/2C east coast gas reserves.[19] Australia is said to have an estimated 186,235 PJ of identified conventional gas and 79,583 PJ of identified CSG.[20] Natural gas production in Australia, including liquefied natural gas (LNG), is projected to grow from 3,109 PJ in 2014–15 to 7,398 PJ in 2049–50, at an annual average rate of 2.5% over the outlook period.[21] Unconventional gas has become important to Australia's energy supply mix as a means of reducing greenhouse gas emissions, ensuring energy security (especially in the east coast markets) and earning export revenue. Production rates continue to increase, as described in the graph below which outlines the CSG production and share of total gas consumption in Australia from 2004–05 to 2014–15.

In Queensland alone, CSG production has increased significantly over the past 15 years as indicated by the graph below.

Given the ever-evolving knowledge base for unconventional gas, particularly tight and shale gas, the outlook for Australian unconventional gas is difficult to define. This position will be improved by increased availability of further exploration drilling and production data. In spite of this, CSG has nevertheless been predicted to continue to be an important sector of the unconventional gas industry with a potential for there to be around 130 years of CSG reserves based on current production levels.[22] With the ongoing development and exploration of unconventional gas resources, Australia's several basins will emerge as having substantial CSG, tight and shale gas resources.

The following sections outline key statistics and proposed CSG and shale gas/ tight gas projects.

3.2 Australian CSG landscape

CSG exploration in Australia has increased substantially over the last 15 years, particularly in Queensland and New South Wales. This has been attributed to the

---

15 *Ibid.*
16 *Ibid.*
17 *Ibid.*
18 "Select Committee on Unconventional Gas Mining – Submission", Australian Petroleum Production & Exploration Association, (March 2016) pp 16-19 www.aph.gov.au/Parliamentary_Business/Committees/Senate/Gasmining/Gasmining/Submissions.
19 *Ibid* at 18.
20 "Australian Energy Resources Assessment: Gas – Summary", Geoscience Australia, (2014) www.ga.gov.au/aera/gas.
21 "Australian Energy Projections to 2049–50", Bureau of Resources and Energy Economics, (November 2014) 44 https://industry.gov.au/Office-of-the-Chief-Economist/Publications/Documents/aep/aep-2014-v2.pdf.

Australian gas production, by gas type

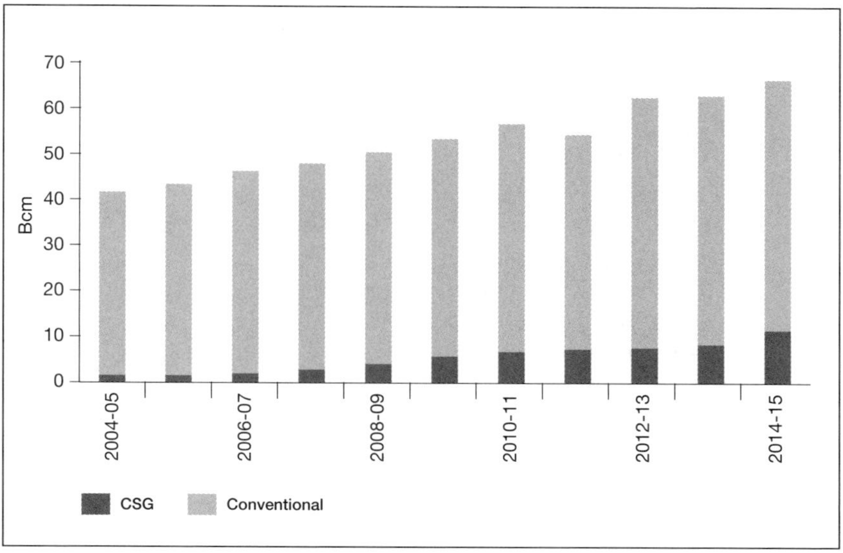

Source: Department of Industry, Innovation and Science (2015) Resources and Energy Quarterly September; Australian Energy Statistics, Queensland Government, Office of the Chief Economist

Queensland CSG Production

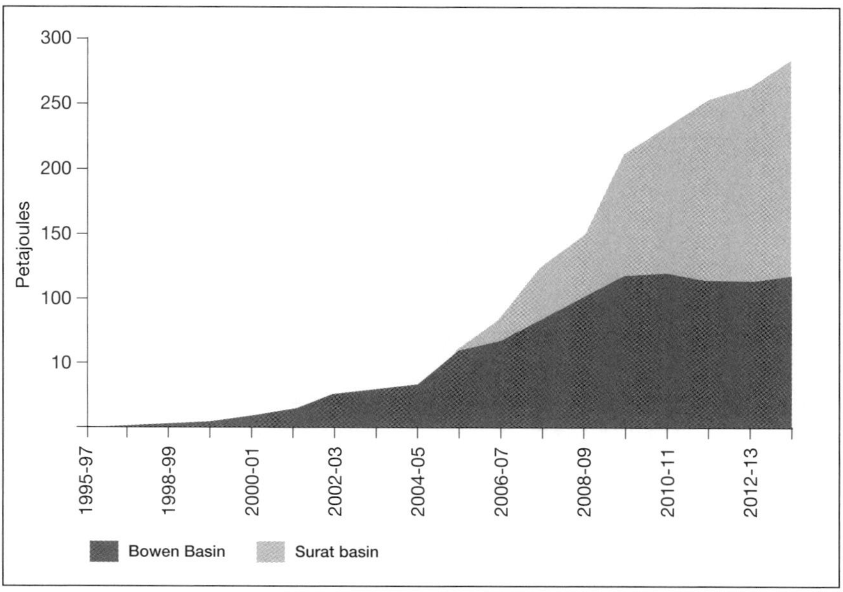

Source: Queensland Government, Office of the Chief Economist

successful development of CSG production in Queensland. Australia's main CSG basins are:

- the Surat, Bowen and Galilee basins in Queensland;
- the Clarence-Moreton Basin in south east Queensland and north east New South Wales; and
- the Sydney, Gunnedah and Gloucester basins in New South Wales.

Almost all of Australia's current CSG reserves are contained in the Surat and Bowen basins in Queensland, and together, these two basins account for 90% of gas production in Queensland.[23]

Given the vast amount of CSG resources in Queensland basins, it is no surprise that Queensland has been the leader in the Australian CSG industry, with a robust and detailed regulatory regime. In fact, the first commercial production of CSG in Australia was in the Bowen Basin from the Dawson River CSG area in 199i6.[24]

In recent years, the CSG industry has been transformed from a solely domestic industry to having a major part to play in the LNG export industry.

The following LNG CSG export projects are currently operating in Queensland:

Project	Participants	Basin	Capacity	First cargo
Australia Pacific LNG	Origin (37.5%) ConocoPhillips (37.5%) Sinopec (25%)	Surat and Bowen basins, Queensland	2 trains with a combined capacity to produce 9 million metric tonnes per annum (MMtpa)[25]	2016
Gladstone LNG	Santos (30%) PETRONAS (27.5%) Total (27.5%) KOGAS (15%)	Surat and Bowen basins, Queensland	2 trains with a combined capacity to produce 7.8 MMtpa[26]	2015

continued on next page

22 "Australian Energy Resources Assessment", Geoscience Australia, (2014) www.ga.gov.au/aera/gas.

23 "Queensland's unconventional petroleum potential: Shale oil and gas, tight gas and coal seam gas", Queensland Department of Natural Resources and Mines, (January 2016) https://www.dnrm.qld.gov.au/__data/assets/pdf_file/0018/291330/unconventional-petroleum-potential.pdf

24 "Coal Seam Gas – Estimation and Reporting of Greenhouse Gas Emissions", Australian Government, www.environment.gov.au/system/files/pages/63b569ff-ae63-4d7b-be54-16f2e79900e0/files/nga-factsheet7.pdf.

25 "Australia Pacific LNG exports first LNG from Curtis Island", Australia Pacific LNG Media Release (11 January 2016), https://www.aplng.com.au/content/dam/aplng/media-release/2016/11012016 APLNG First Export Media Release FINAL.pdf.

26 "The Project", Santos, www.santosglng.com/the-project.aspx.

Project	Participants	Basin	Capacity	First cargo
Queensland Curtis LNG	QGC (wholly owned subsidiary of Royal Dutch Shell)	Surat Basin, Queensland	2 trains with a combined capacity to produce 8.5 MMtpa[27]	2014

3.3 Australian shale gas and tight gas landscape

Australia has significant shale gas resources, but these are poorly quantified in contrast to CSG and conventional gas resources. There is an estimated 619 trillion cubic feet (tcf) of prospective recoverable resources, being 681,273 PJ.[28] Likely shale gas formations have been identified in many basins, including the Cooper, Perth, Amadeus, Canning, Georgina and McArthur basins.

Australia does not currently have any reserves of tight gas, but reports have estimated that Australia has 44 tcf or 48,894 PJ of prospective tight gas resources.[29] The two basins containing the largest resources of tight gas are the Perth Basin, which has an estimated 8.4 tcf or 9,300 PJ[30] of tight gas and the Cooper Basin, which contains about 9.5 tcf or 10,500 PJ[31] of tight gas. The Gippsland Basin has been estimated to contain about 0.43 tcf or 500 PJ of tight gas.[32] These occurrences of tight gas are located close to infrastructure and have been considered for commercial production. Other occurrences are located in basins without infrastructure and are currently not considered for commercial production.

In contrast to the United States, where shale and tight gas provides for around 48% of total dry natural gas production,[33] the commercial production of shale gas in Australia is still in its infancy. Most projects are at an early exploration or appraisal phase.

Australian Securities Exchange (ASX)-listed companies that are active in the shale gas hunt include:

- Buru Energy, which is jointly exploring with Japan's Mitsubishi for oil and gas in the Canning Superbasin, northwest of Western Australia, with an aim to deliver 1,500 PJ of gas into the Western Australia domestic market;
- Beach Energy, which has been exploring for shale gas in its Cooper Basin tenements. Beach recently merged with Drillsearch, which has also been active in exploring and developing shale gas resources in the Cooper Basin;
- New Standard Energy, which is spending A$110 million to explore for gas in the Canning Basin; and

27 "Factsheet Australia Profile", BG Group, (August 2015) www.bg-group.com/files/pdf/qgc/2481_qgc-bg_ausprofile.webfinal.pdf.

28 "Australian Energy Resources Assessment: Gas – Summary", Geoscience Australia, (2014) www.ga.gov.au/aera/gas.

29 *Ibid.*

30 *Ibid.*

31 *Ibid.*

32 *Ibid.*

33 "How much shale gas is produced in the United States?", EIA, (14 June 2016) www.eia.gov/tools/faqs/faq.cfm?id=907&t=8.

- AWE, which has commenced testing for shale gas in the reportedly shale-rich Perth Basin.[34]

Other ASX-listed companies that are also active in the Australian shale gas industry include: Santos, Senex Energy, Cooper Energy, Norwest Energy, Empire Oil and Gas, and Icon Energy.

The success of the CSG LNG projects in Queensland and strong gas prices encouraged international majors' investment in Australia's budding shale and tight gas industries. However, from 2014–15, this enthusiasm started to wane as declining commodity prices led to cost pressures. Australia's not-yet-proven shale and tight gas projects soon became 'non-core' and expendable assets. For example, Chevron elected not to proceed to stage two of its Cooper Basin shale gas exploration joint venture with Beach Energy, after spending A$190 million in the first stage. This decision was made notwithstanding the positive drilling results from the first stage. US-based Hess Corporation forfeited its rights in a shale gas exploration tenement in the Beetaloo Basin in the Northern Territory after it failed to elect to commit to further drilling.[35]

4. Australian regulatory framework

4.1 Australian unconventional gas regulation

Unconventional gas in Australia is regulated at a state and federal level. Since Australia's unconventional gas reserves are found in several states, different regulations apply depending upon where the resource is located.

The state and territory governments have independent legislative power in all areas not specifically designated to the Federal Government under the Australian Constitution, and where there is an inconsistency between state and federal laws, federal laws will prevail.

As such, each state has its own legislation to deal with applying for and development of mining and petroleum titles. While there have been calls to harmonise the regimes across Australia, this has not occurred. The regulatory regimes are at different stages of development based on the stage of development of the industry for that state.

The Australian Petroleum Production & Exploration Association has estimated that an onshore gas company operating nationally needs to comply with over 70 pieces of state and federal legislation, as well as numerous additional regulations, operational policies and local council regulations.[36] This demonstrates the significant compliance costs faced by companies when operating in this space. Below is a high-level description of the current regimes in Queensland, New South Wales, South

---

34 "About AWE", AWE Limited, www.awexplore.com/irm/content/about-awe1.aspx?RID=319.

35 "Falcon Retains Interest in Beetaloo Permits, Northern Australia, Hess Request to Defer Drilling Decision Rejected by Falcon Board", Falcon Oil & Gas, (1 July 2013) www.falconaustralia.com.au/uploads/pdf/13-07-01_Non-election_by_Hess.pdf.

36 "Select Committee on Unconventional Gas Mining Submission", Australian Petroleum Production & Exploration Association Limited, (March 2016) www.appea.com.au/wp-content/uploads/2016/04/APPEA-Submission-Lazarus-Inquiry-final.pdf.

Australia, Western Australia, Victoria and the Northern Territory, and some of the particular challenges those states are facing with respect to legislating for the growing unconventional gas industry.

4.2 Queensland

The Queensland petroleum and gas industry is currently regulated by a series of legislative instruments.

All petroleum titles are granted and regulated under either the Petroleum and Gas (Production and Safety) Act 2004 (Qld) (for recent applications) or the Petroleum Act 1923 (Qld) (for older applications). Land access and other elements are also dealt with through the Mineral and Energy Resources (Common Provisions) Act 2014 (Qld), which commenced on 26 September 2016.

The Petroleum and Gas (Production and Safety) Act was enacted in 2004, at a time when CSG accounted for more than 27% of Queensland's total supplied gas in that year.[37]

With the aim of promoting growth within the state's vital petroleum and gas industries, and to optimise the utilisation of its CSG resources, the Petroleum and Gas (Production and Safety) Act implemented a new CSG regime providing mechanisms for the efficient and effective management of Queensland's CSG resources.

The objectives of the regime as stated in the explanatory memorandum are to:

- provide clarity of rights and tenure;
- deal with safety issues associated with the exploration and production of CSG; and
- ensure that the future safe and efficient mining of coal seams is not hindered by petroleum exploration and production activities.[38]

Through the new regime, the Petroleum and Gas (Production and Safety) Act defines the legal rights, obligations and priorities for developing co-existing petroleum, coal and CSG resources. The objective is to create greater certainty for potential explorers and developers investing in the industry. CSG has been defined in the Petroleum and Gas (Production and Safety) Act as petroleum (in any state) occurring naturally in association with coal or oil shale, or in strata associated with coal or oil shale mining.

As is the case with most minerals, CSG is deemed under the Petroleum and Gas (Production and Safety) Act as the property of the state.[39] Although previously allowed, the law does not allow for commercial production of CSG under mining tenements. Rather, only incidental extraction of CSG is permitted under a coal or oil shale mining tenement in Queensland.[40] In order to be allowed to commercially produce CSG, the mining company must obtain a petroleum licence in accordance

---

37 "Parliamentary Debates", Queensland, Legislative Assembly, 12 May 2004 at 897.
38 Explanatory Notes, Petroleum and Gas (Production and Safety) Bill 2004 (Qld) 1.
39 Petroleum and Gas (Production and Safety) Act 2004 (Qld) s 26.
40 Mineral Resources Act 1989 (Qld), Part 7AA. Section 318CM of the Act sets out a limited entitlement to mine CSG under a mining lease.

with either the Petroleum and Gas (Production and Safety) Act[41] or the Petroleum Act. The process for obtaining this licence, in pursuit of the Petroleum and Gas (Production and Safety) Act's underlying objectives, requires the prospective developer to pay due regard to the protection of the interests of existing tenement holders and the future development of the resources within those tenements.[42]

Before a company is permitted to commence development of a CSG project, it must first obtain certain environmental approvals. An application must be made for the required environmental authority under the Environmental Protection Act 1994 (Qld). As part of the approval process, an environmental impact statement (EIS) may either be required by the criteria set out by the Queensland Department of Environment and Heritage Protection and the Environmental Protection Act, or can be voluntarily completed to support an application. The proposed CSG project is then classified based on its risk of environmental harm, with procedures and requirements imposed in accordance with its classification.[43] EIS proposals that involve high risks will continue to require closer scrutiny by the Department of Environment and Heritage Protection, which may apply site-specific conditions where appropriate (resource projects triggering the EIS process will attract the most intensive scrutiny).[44]

For certain projects, the EIS process requires the company to submit an environmental management plan addressing the environmental values that are likely to be affected and the impact that the proposed project may have on those values, and proposing environmental protection commitments to protect those values. The plan should ideally assist the Department of Environment and Heritage Protection with developing conditions around the project.[45] It should be added that despite CSG water being treated as waste under the Environmental Protection Act, it may be used as a resource by the relevant company if it is deemed to have a beneficial use.[46] The purposes of the EIS and the EIS process are to: assess the potential adverse and beneficial environmental, economic and social impacts of the project; assess the measures proposed to minimise adverse environmental impacts of the project; promote transparency about these issues; and provide information to the commonwealth and state authorities to make informed decisions.

A CSG compliance unit was established in 2011, and which now forms part of the Queensland Department of Natural Resources and Mines. This is a multi-disciplinary group that incorporates experts from all relevant fields including petroleum, groundwater and gas safety fields. The compliance unit's role is to monitor CSG operations for compliance with the laws and policies of Queensland.

---

41 Petroleum and Gas (Production and Safety) Act 2004 (Qld) s 800.

42 Under the Petroleum and Gas (Production and Safety) Act Ch 3, Pt 6, an application must be accompanied by a CSG statement and other CSG assessment criteria (including the safety requirements under Chapter 9 of the act, the initial development play requirements and the public interest) that address these issues.

43 N Swayne, "Regulating coal seam gas in Queensland: lessons in an adaptive environmental management approach?" (2012) 29 *Environmental and Planning Law Journal* 63 at 185.

44 "Environmental Impact Statement (EIS) process", Queensland Government, (18 October 2016) https://www.ehp.qld.gov.au/management/impact-assessment/eis-processes/.

45 *Ibid.*

46 *Ibid.*

Additionally, the compliance unit is designed to be the point of call for queries and complaints regarding land access and environmental concerns, and to manage and investigate all complaints and media inquiries. The introduction of the compliance unit was in response to the growing scale of the CSG industry in Queensland. Western Australia and New South Wales' respective CSG industries are not quite of the scale of Queensland's and as such have not introduced similar groups at this stage.

From a holistic point of view, the development of the CSG industry in Queensland has been guided by an adaptive management approach. This approach has been incorporated in the legislation regulating the industry, including the Petroleum and Gas (Production and Safety) Act and the Environmental Protection Act. According to the Queensland Government, the approach is "a system to monitor and instigate change where required" as "technologies develop over time".[47]

In addition, the Queensland Government has established the GasFields Commission as an independent statutory body under the GasFields Commission Act 2013 (Qld) to assist with the coexistence of landowners, regional communities and the onshore gas industry. The GasFields Commission was intended to give communities more power regarding the development of CSG – for example, by holding stakeholder meetings, actioning queries from landowners and providing accurate information about the industry to the public.[48]

4.3 New South Wales

The New South Wales unconventional gas industry is currently regulated in the same way as conventional onshore petroleum activity. The legislation relating to CSG is the Petroleum (Onshore) Act 1991 (NSW) and the Environmental Planning and Assessment Act 1979 (NSW).

Before undertaking any CSG-related activities, companies must first apply to the minister for industry, resources and energy for the relevant licence under the Petroleum (Onshore) Act. Under this act, the main licences required to prospect for and extract CSG are exploration licences, petroleum assessment leases and petroleum production leases.

Whilst not as extensive as the CSG regime in Queensland, the Petroleum (Onshore) Act nevertheless seeks to maintain the protection of the environment as one of its underlying objectives. For example, in deciding whether to grant the licence, the minister must take into account the impact of the project on the environment or on the land over which the title is sought.[49] The minister may grant the licence subject to any conditions regarding the environmental management, protection and rehabilitation of the land and the safety of the public in relation to prospecting and mining operations.[50]

---

47 "Adaptive management", Queensland Government Department of Environment and Heritage Protection, (1 April 2013) https://www.ehp.qld.gov.au/management/non-mining/adaptive-management.html.

48 "Annual Report 2015-16", GasFields Commission, (13 September 2016) 8-9 www.gasfieldscommissionqld.org.au/resources/gasfields/annual-report-2015-16.pdf.

49 Petroleum (Onshore) Act 1991 (NSW), Sch 1B, Pt 2.

50 *Ibid.*

The majority of petroleum activities need development approval or approval from the New South Wales Department of Industry, Skills and Regional Development before commencement.[51] The department is predominantly responsible for the administration of the Petroleum (Onshore) Act, including: the granting, renewal and administration of all permits and titles; the environmental assessments of exploratory activities; setting standards for safety requirements; and management of rehabilitation efforts, among other responsibilities.[52]

Environmental assessment is required as part of the approval process, pursuant to the Environmental Planning and Assessment Act. The Environment Protection Agency is responsible for compliance and enforcement of conditions of approval for CSG activities in New South Wales. Companies engaging in CSG activities must hold an environmental protection licence issued by the Environment Protection Agency and also comply with the extensive provisions under the Protection of the Environment Operations Act 1997 (NSW).

Under the Environmental Planning and Assessment Act, CSG operations will require additional approval from the minister for lands and water pursuant to the Water Management Act 2000 (NSW) if their activities will involve penetration, interference, obstruction of the flow, the taking of water from an aquifer in the course of carrying out mining, or disposal of water in an aquifer.[53]

The use of BTEX chemicals (benzene, toluene, ethylbenzene and xylenes) in the fracking process is currently banned in New South Wales.[54] All CSG drilling additives and CSG fracking additives must be tested by a National Association of Testing Authorities-certified laboratory and demonstrated to meet Australian drinking water health guideline values.[55]

There has been much debate in New South Wales about its growing CSG industry. In July 2011, the New South Wales Government instigated a moratorium on the use of fracking in the extraction process, in response to community concern over the potential for pollution of aquifers. The primary impetus behind the moratorium was the concern over the use of certain chemicals in the fracking process. The moratorium was eventually lifted and the government now assesses each CSG project on a case-by-case basis.

Notwithstanding the lifting of the moratorium, the regulatory environment in New South Wales remains challenging for unconventional gas developments. The state government had instituted a CSG buy-back programme. In 2015, the government released a gas plan which calls for a reset of the application system whereby all existing applications at that time were expunged and a new system was implemented. As part of the new system, CSG exploration licences are to be limited to 15% of the state (down from 60%).

---

51 "Exploration licences and regulation", New South Wales Government, Department of Industry, Skills and Regional Development, www.resourcesandenergy.nsw.gov.au/landholders-and-community/minerals-and-coal/exploration, viewed 30 November 2016.

52 *Ibid.*

53 Water Management Act 2000 (NSW) s 91L.

54 "Ban on use of BTEX compounds in CSG activities", New South Wales Department of Industry, www.industry.nsw.gov.au/policies/items/ban-on-use-of-btex-compounds-in-csg-activities.

55 *Ibid.*

4.4 South Australia

In South Australia, all operations associated with upstream petroleum operations including unconventional gas such as CSG, shale and tight gas are regulated by the Petroleum and Geothermal Energy Act 2000 (SA) and the Petroleum and Geothermal Energy Regulations 2013 (SA). The Petroleum and Geothermal Energy Act and the regulations are administered by the Energy Resources Division of the Department of State Development. The key aims of the regulatory regime are to minimise and manage the risks that petroleum projects may pose to the health and safety of the community as well as the environment.

The Petroleum and Geothermal Energy Act provides that regulated activities under the Act, such as the production of CSG, may not be carried out unless there is an approved statement of environmental objectives in place based on an environmental impact report.[56] Such report must: take into account cultural, amenity and other values of Aboriginal and other Australians insofar as those values are relevant to the assessment; take into account risks inherent in the regulated activities to the health and safety of the public; and contain sufficient information to make possible an informed assessment of the likely impact of the activities on the environment.

These documents are put in place to identify all potential risks related to the activity and appropriate risk mitigation strategies, as well as the environmental objectives to be achieved in addressing these risks.

In December 2012, the South Australian Government published a Roadmap for Unconventional Gas Projects in South Australia. The roadmap was initially produced to inform stakeholders and set the course for environmentally sustainable development of South Australia's large endowment of unconventional gas.

The roadmap highlights the effectiveness of the Petroleum and Geothermal Energy Act through its one-stop-shop (through the Energy Resources Division of the Department of State Development) approach to the approval process for exploration and production operations, as well as petroleum transmission pipelines, citing the Australian Productivity Commission which recognised South Australia's approach as a model for other jurisdictions to emulate.[57]

The Petroleum and Geothermal Energy Act also ensures that through compliance with its objectives, upstream petroleum operations will also be compliant with the relevant Commonwealth legislation, such as the Native Title Act 1993 (Cth) and the Environment Protection and Biodiversity Conservation Act 1999 (Cth).[58]

Areas of focus as outlined by the roadmap include ensuring that regulators have relevant and up-to-date capabilities and that there exists an effective management system for risks associated with regulatory capture and environmental harm. The Energy Resources Division of the Department of State Development has been

---

56 Petroleum and Geothermal Energy Act 2000 (SA) ss 96 and 97.

57 "Roadmap for Unconventional Gas Projects in South Australia", Division of the Department for Manufacturing, Innovation, Trade, Resources and Energy, (December 2012) www.statedevelopment.sa.gov.au/upload/mineral-and-energy-resources/Roadmap_Unconventional_Gas_Projects_SA_12-12-12.pdf at p135, viewed 1 December 2016.

58 *Ibid.*

holding regular roundtables to monitor the progress in implementing the 125 recommendations from the roadmap.[59]

4.5 Victoria

In Victoria, CSG exploration and extraction is regulated under the Mineral Resources (Sustainable Development) Act 1990 (Vic) and tight, shale and conventional gas under the Petroleum Act 1998 (Vic). All applications are also subject to regulation under a range of other acts such as the Water Act 1989 (Vic) and the Environment Protection Act 1978 (Vic).

In response to the recent 2015 Parliamentary Inquiry into Onshore Unconventional Gas in Victoria, the Victorian Government has announced a permanent ban on all unconventional gas exploration and development, as well as a moratorium on conventional onshore gas development. This ban is currently contained in the Resources Legislation Amendment (Fracking Ban) Bill 2016 which is before the Victorian Parliament.

At the time of publishing, legal action by way of judicial review has been initiated by Lakes Oil in the Supreme Court of Victoria challenging the ban.[60]

4.6 Northern Territory

All exploration and extraction activities in the Northern Territory are governed primarily by the Mining Management Act (NT) and the Petroleum Act (NT). A raft of other legislation is also relevant to any project, such as the Aboriginal Land Rights Act and the Native Title Act, which are important for land access agreements.

However, the legislative regime has been temporarily displaced by the new Northern Territory Government, elected in August 2016, which vowed to place a moratorium on all unconventional onshore gas activities in the territory. This moratorium was implemented on 14 September 2016 as a moratorium on hydraulic fracturing of unconventional gas reservoirs, including for exploration, extraction and production. It includes diagnostic fracture injection testing. The government has said the moratorium will be in place until the completion of a Fracking Inquiry, and at that time, the government will make a decision on whether to lift the ban and under what conditions. The inquiry is expected to take up to four years to complete.

4.7 Western Australia

In response to the heightened level of community interest in CSG arising in eastern Australia relating to hydraulic fracturing, the Western Australian Government has committed to a regulatory regime that ensures responsible development of unconventional gas resources while protecting the environment and public health.

Onshore extraction of oil and gas is regulated by the Petroleum and Geothermal

---

59 "Petroleum", Government of South Australia, Department of State Development, (30 October 2015) http://petroleum.statedevelopment.sa.gov.au/old-site/prospectivity/basin_and_province_information/unconventional_gas/unconventional_gas_interest_group/roadmap_for_unconventional_gas_projects_in_sa/roundtable_for_oil_and_gas_projects_25_september_2015.

60 "Victoria proposes frac ban law for CSG, tight oil, gas", *Oil & Gas Journal*, (22 November 2016) www.ogj.com/articles/2016/11/victoria-introduces-legislation-to-ban-gas-fracing.html.

Energy Resources Act 1967 (WA), the Petroleum and Geothermal Energy Resources (Resource Management and Administration) Regulations 2015, and the Petroleum and Geothermal Energy Resources (Environment) Regulations 2012.

The Petroleum and Geothermal Energy Resources Act regulates projects from the exploratory stage to the end of the life of the project, and sets out the requirements for the drilling of wells, the grant of production licences and the registration of titles.

All exploration and extraction activities require an approved environmental plan to be in place, and it is an offence to conduct an exploration or extractive activity without such a plan.[61]

Where a proponent intends to carry out drilling and fracking operations as part of its particular project, the Department of Mines and Petroleum requires full disclosure of the purported chemicals in both the drilling application and the environmental management plan, as well as detailed information about the fracturing programme in order to determine its acceptability.[62] The Department of Mines and Petroleum has the right not to grant approval where potentially harmful chemicals are going to be used.

4.8 Tasmania

CSG exploration and extraction in Tasmania is primarily governed under the Mineral Resources Development Act 1995 (Tas) which broadly follows the requirements in other states for obtaining permits.

In March 2014, the Tasmanian Government announced a moratorium on hydraulic fracking in the state. This moratorium has been extended to March 2020.[63] The justification for the moratorium is to protect Tasmania's prime agricultural land until a proper inquiry into hydraulic fracking can be concluded. There had been no hydraulic fracking in Tasmania prior to the moratorium, although a small number of exploration permits had been granted. Under the moratorium, permit holders will continue to be able to explore for gas; however, they will not be able to use any methods involving hydraulic fracking to do so.

4.9 Federal environmental approvals

As previously noted, the regulation of the CSG industry is primarily the responsibility of state and territory governments. However, if a proposed CSG activity is likely to impact on a matter of national environmental significance, then approvals may need to be obtained from the commonwealth minister for the environment. Even if a CSG proposal has an environmental impact, the environment minister may not intervene if the CSG proposal is not likely to have a significant impact on a matter of national environmental significance. This is not due to the unimportance of these other matters but rather the arrangement under the Australian Constitution, which grants the primary responsibility for environmental protection to the states. The legislation

61 Petroleum and Geothermal Energy Resources (Environment) Regulations 2012 (WA) reg 6.

62 "Hydraulic fracturing of gas reserves", Environmental Protection Authority WA, EPB No 22 (December 2014) http://edit.epa.wa.gov.au/EPADocLib/EPB22-Fracking-171214.pdf.

63 Premier of Tasmania, "Fracking Moratorium to Continue", (26 February 2016) www.premier.tas.gov.au/releases/fracking.

governing the Federal Government's right to intervene in this regard is the Environment Protection and Biodiversity Conservation Act 1999 (Cth).

If an activity is likely to have a significant impact on any of the nine currently listed matters of national environmental significance, an environmental assessment is to be carried out by the environment minister before a determination is made.[64] Of noteworthy importance is the bilateral agreement that exists between New South Wales and the Commonwealth permitting the environment minister to rely on an assessment carried out under the Environmental Planning and Assessment Act. Importantly, in 2013, the Commonwealth amended the Environment Protection and Biodiversity Conservation Act to include water resources as a matter of national environmental significance when affected by coal mining or CSG extraction, and this allows the Commonwealth to comprehensively assess any CSG activities that will have a significant impact on water resources.[65]

4.10 National partnership agreement for regulation of CSG

The National Partnership Agreement on Coal Seam Gas and Large Coal Mining Developments came into effect on 14 February 2012 and has been adopted by the Australian Federal Government, Queensland, New South Wales, South Australia, Victoria and the Northern Territory.

By adopting the partnership agreement, the federal, state and territory governments agree to: address public concerns about the actual and potential impacts of CSG on water resources by substantially improving scientific evidence that informs regulatory decisions; enhance the transparency and openness of regulatory processes; and ensure that expert advice on all CSG proposals are available to communities, governments and the industry.[66] The partnership agreement represents recognition by the federal, state and territory governments that there exists a mutual interest in the long-term health, quality and viability of Australia's water resources and the sustainable development of CSG and coal mining industries in Australia.[67]

In particular, the partnership agreement aims to:

- provide governments with expert scientific advice on CSG and large coal mining development proposals that are likely to have a significant impact on water resources;
- commission public research and bioregional assessments in areas of high CSG and/or large coal mining developments in order to improve scientific knowledge and understanding of the impacts of these developments, both in specific locations and on water resources more broadly; and
- make public its advice and the findings of all research.[68]

---

64 "What is protected under the EPBC Act?", Australian Government Department of the Environment and Energy, https://www.environment.gov.au/epbc/what-is-protected.

65 "Coal and coal seam gas regulation", Australian Government, Department of the Environment and Energy, https://www.environment.gov.au/water/coal-and-coal-seam-gas/regulation.

66 COAG, National Partnership Agreement on Coal Seam Gas and Large Coal Mining Development, cl 4 www.federalfinancialrelations.gov.au/content/npa/environment/national-partnership/past/coal_mining_development_NP.pdf.

67 *Ibid* at cl 3.

68 *Ibid* at cl 11.

Under the partnership agreement, the Commonwealth agrees, amongst other things, to provide financial contribution to the states and territories to support the implementation of the agreement and to monitor and assess the delivery of actions under it. The partnership agreement requires the Commonwealth to establish an independent expert scientific committee, the purpose of which is to provide robust scientific advice to governments on CSG and large coal mining developments which are likely to have a significant impact on water resources.[69] The states and territories agree to: provide input to the committee's research agenda; seek advice from the committee at appropriate stages of the approvals process for CSG proposals; and amend laws and regulations as necessary to ensure that CSG developments that are likely to have a significant impact on water resources are referred to the committee for advice.[70]

5. Current issues in unconventional gas

Many of the current issues in unconventional gas in Australia relate primarily to the CSG industry. This reflects the fact that the Australian CSG industry is more mature than the shale gas industry, and as projects commence development they are subject to increased scrutiny by the community, government and affected stakeholders.

In addition, as many of the large-scale CSG LNG export projects took final investment decisions, parallels were drawn with the environmental and social impacts seen in the shale gas industry in the United States. The exploitation in the United States of shale gas and in Australia of CSG were, and still are by some sections of Australian communities, seen as indistinguishable in the absence of information about the CSG industry.

5.1 Competing land use

Unlike conventional gas resources, developing and exploiting CSG resources requires a large number of wells, rigs and collection and transmission pipelines. Access to land is a critical component of CSG developments. Australia's CSG-producing basins are located in a range of areas including cities, agricultural regions and coal mining areas. As such, CSG projects must compete for land with other resource projects, farmers and local communities.

The CSG industry has been subject to significant public opposition from groups such as communities on the north coast of New South Wales and the "Lock the Gate Alliance" with its call to farmers to lock the gate to prevent CSG companies from accessing agricultural lands. The challenge for governments is to put in place a regulatory regime which balances the needs of landholders and the desire to exploit natural resources in a safe and effective manner.

It is instructive to look at the experience in Queensland and New South Wales to see how this issue is being managed.

---

69 *Ibid* at cl 14.
70 *Ibid* at cl 15.

(a) *Queensland*

Under the Queensland legislative regime, the holder of a petroleum authority must ensure that it carries out its authorised activities in a way that does not unreasonably interfere with anyone else carrying out a lawful activity.[71] This includes other authority holders in the area as well as local landholders carrying out lawful activities. The regime provides that where an activity interferes with any lawful operations being carried out on the land, the Land Court may order changes in the activity to remove such interference. CSG projects will necessarily require the construction of key infrastructure on the land such as wellheads, compressor stations, treatment facilities and access roads.[72] The mere existence of this infrastructure in addition to the noise potentially created by the development of such CSG projects would be highly likely to disrupt the lawful practices carried out by a local landholder.[73]

Taking effect from October 2010, the Queensland Government introduced a uniform Land Access Code in an attempt to balance the interests of the agricultural and resources sector, applying to all major resources projects including petroleum projects. The code contains best practice guidelines for landholders and companies about how to manage consultation and compensation processes, as well as mandatory conduct conditions with which companies must comply when undertaking unauthorised activities on private land. The code was modified slightly in 2016 to reflect administrative and legislative changes (eg, to certain definitions); however, the intent, scope and operation of the code have remained largely unchanged.

Queensland's competing land use issues have also been addressed by the introduction of the Regional Planning Interests Act 2014 (Qld) which came into force in 2014 replacing the Strategic Cropping Land Act 2011 (Qld). The Regional Planning Interests Act requires land which is an "area of regional interest" to undergo a "regional interests development approval", unless it is an existing resources activity. 'Regional interest areas' are areas which are priority agricultural areas, priority living areas, strategic cropping land areas and strategic environmental areas.[74] There are a number of exemptions to the additional approval process under the Regional Planning Interests Act, such as if a compensation agreement exists between the landowner and authority holder, and the activity is unlikely to have a significant impact on the land, a priority agricultural area or a strategic cropping area.[75]

(b) *New South Wales*

The law in New South Wales requires prospecting licence applicants to enter into an access arrangement with the owner or the occupier of the land they wish to enter, where it is appropriate.[76] The applicant will not be permitted to enter the land until

71 Petroleum and Gas (Production and Safety) Act 2004 (Qld) s 804.
72 See N Swayne, footnote 44 at 172.
73 *Ibid.*
74 Regional Planning Interests Act 2014 (Qld) s 7.
75 *Ibid*, s 22.
76 Petroleum (Onshore) Act 1991 (NSW) Pt 4A.

such a time as this agreement has been reached. However, if the landholder is uncooperative in entering into this agreement, the applicant may request the access arrangement be determined through arbitration or other prescribed methods. The landholder is entitled to compensation where its land is, or is likely to be, injuriously affected by the CSG operations.[77]

In May 2011, the New South Wales Government put in place a moratorium for 60 days over new CSG exploration licences, in a bid to resolve some ongoing issues surrounding land use and CSG mining.[78] In an attempt to resolve these issues, a parliamentary inquiry, as mentioned in Section 4.3 of this chapter, was conducted to explore and report on the environmental, economic and social impacts of CSG activities.

(c) *Liability for damage to land*

Farmers are not only concerned about damage to aquifers relied on for agricultural production, they are also worried about the 'footprint' left by CSG developments on the agricultural productivity of their land. The relevant state legislation governing the carrying out of CSG activities provides compensation mechanisms for landholders who have suffered damage to or a diminution in value of their land as a result of a petroleum authority holder's activities. Generally, a licensee or lessee must compensate each owner or occupier of private land for any loss or damage arising from the carrying out of activities under the licence or lease on the land.[79] A landholder and licensee or lessee must enter into an agreement about how the compensation is to be determined, failure of which would require the parties to refer the matter to an independent third party for determination.[80]

5.2 Water resource management

Community and landholders, especially farmers, have expressed strong concerns about the potential adverse effects of CSG production on Australia's aquifers, especially in agricultural production areas, and the disposal of wastewater extracted through the production of CSG.

As illustrated by the following map, Australia's CSG-producing basins overlap with one of the world's largest groundwater aquifers, the Great Artesian Basin. The basin is approximately 1.7 million km^2[81] and provides a freshwater source for a number of uses, including watering of livestock, irrigation, mining, and industrial and domestic purposes.

Therefore, interference with the quality of the water within the basin would

---

77 *Ibid*, 107(1).

78 "Inquiry calls for freeze on coal seam gas production", *Sydney Morning Herald*, (1 May 2012) www.smh.com.au/environment/inquiry-calls-for-freeze-on-coal-seam-gas-production-20120430-1xv5x.html.

79 Mineral and Energy Resources (Common Provisions) Act 2016 (Qld) Subdivisions 3-5; Petroleum (Onshore) Act 1991 (NSW) ss 107 and 108; Petroleum and Geothermal Energy Resources Act 1967 (WA) ss 17 and 19; and Petroleum and Geothermal Energy Act 2000 (SA) s 63.

80 Mineral and Energy Resources (Common Provisions) Act 2016 (Qld), Sub-divisions 3-5; Petroleum (Onshore) Act 1991 (NSW) s 108(2); Petroleum and Geothermal Energy Resources Act 1967 (WA), ss 17(4) and 19; and Petroleum and Geothermal Energy Act 2000 (SA), s 63(4).

81 "Discover groundwater", Murray-Darling Basin Authority, www.mdba.gov.au/discover-basin/water/discover-groundwater.

Map of the Great Artesian Basin

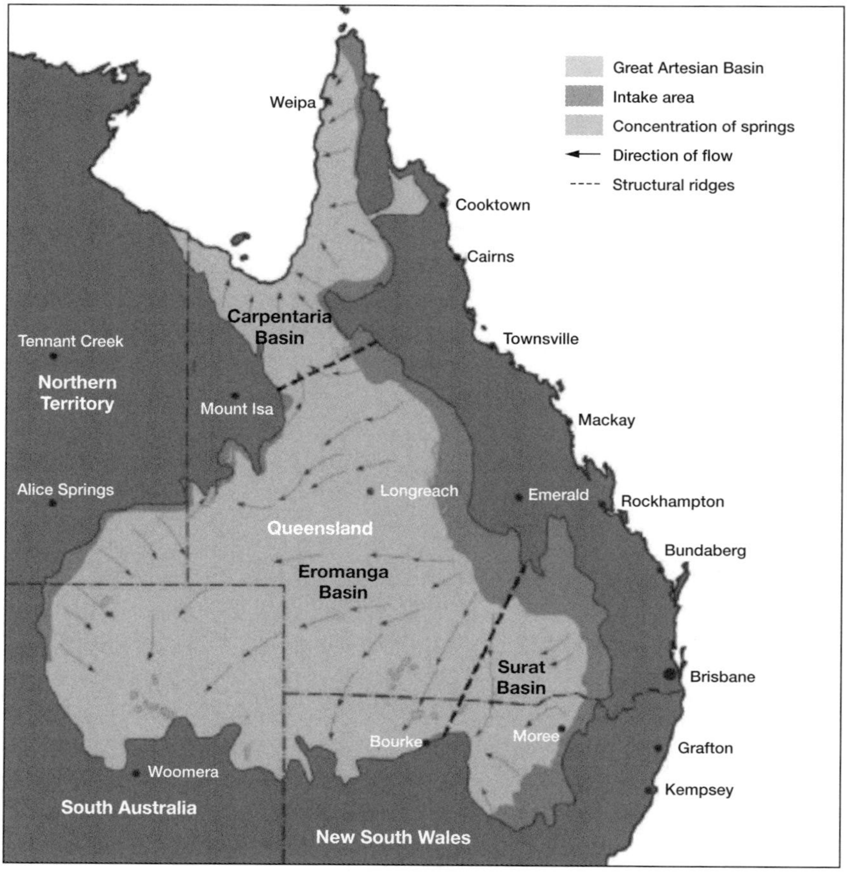

Source: State of Queensland, Department of Environment and Resource Management, Factsheet on Great Artesian Basin

cause problems for a broad range of industries. Management of both the quantity and the quality of the water used in CSG mining has become a significant issue.

As previously mentioned, CSG extraction requires a significant quantity of water to start the extraction process, but the aquifer itself must be de-watered to lower the pressure to allow the gas to escape from the coal seam. While there is debate about the extent of possible impacts of the CSG industry on the aquifer, a concern has been expressed that the volume of water to be removed from the aquifer will lower the pressure of the aquifer, reducing the ability to recover groundwater in the future.

In addition, there are concerns regarding the impact of chemicals used in the CSG extraction process on the aquifer. This arises not only from produced water which may runoff into the aquifer, but also from chemicals used in the fracking process that may contaminate aquifers, particularly given that thousands of wells are drilled to support CSG production.

A statement released by the former National Water Commission[82] identifies and summarises the following areas of potential risk to sustainable water management resulting from CSG activities:

- Extraction of large volumes of water will impact on connected surface and groundwater systems.
- The depressurisation of coal seams could adversely affect other water users and the environment through changes in water availability and reductions in surface water flows in connected systems.
- The production of large volumes of treated wastewater could, if released to surface water systems, alter natural flow patterns and can significantly impact water quality and river and wetland health.
- The practice of fracking has the potential to induce cross-contamination between aquifers, impacting on groundwater quality.
- The reinjection of treated wastewater into the aquifers has potential to change the beneficial use characteristics of those aquifers.[83]

Fortunately, most of the state legislation governing CSG-related activities provides for some form of sustainable water management, requiring disclosure of the possible impact of its proposed CSG-related activities on water resources within the area, as well as an outline of the management systems to be put in place to minimise the risk of harm to the environment. Queensland in particular has imposed 'make good' obligations on CSG companies, requiring the company to take restoration measures to restore an existing water bore's capacity if it has been affected by CSG operations undertaken by the company.[84] As recognised by the former National Water Commission, such an adaptive and precautionary environmental management approach is essential for progressive improvement in the understanding of impacts and cumulative effects of CSG development on water resources,[85] and should be a model for other jurisdictions to follow.

5.3 Overlap of coal mining and CSG activities

In Queensland, there is significant overlap between permits for coal and petroleum exploration and production.[86] In fact it was estimated by the former Department of Employment, Economic Development and Innovation that the extent of the overlap in Queensland's major coal basins is approximately 30% in the Galilee Basin, 50% in the Bowen Basin and 25% in the Surat/Clarence Moreton Basin.[87] The introduction of the Petroleum and Gas (Production and Safety) Act in 2004 was accompanied by a policy for addressing overlapping tenure issues. Under the policy, coal and CSG

---

82 The National Water Commission was abolished in 2015 and its responsibilities transferred to a number of departments and bodies such as the Productivity Commission, the Bureau of Meteorology and the Australian Bureau of Agricultural and Resource Economics and Sciences.

83 "Position Statement: Coal Seam Gas and Water", National Water Commission, www.nwc.gov.au/__data/assets/pdf_file/0003/9723/Coal_Seam_Gas.pdf.

84 Water Act 2000 (Qld) Part 5.

85 National Water Commission, n 91.

86 Plumb J, "Overlapping tenure in Queensland: a new regime", http://gastoday.com.au/news/overlapping_tenure_in_queensland_a_new_regime/056771/.

87 *Ibid.*

proponents were permitted to explore the same area, but where the activities come into conflict, priority is given to the proponent which had commenced activities first. As a result of the booming CSG-to-LNG industry, Queensland has seen an increase in the number of new applications being lodged for petroleum leases as a means of securing long-term tenure.[88] The bulk of these applications are made over coal exploration ground and experience delays to commencement of production. These delays will in turn generate tension between the competing interests of CSG and coal companies which would inevitably lead to a breakdown of negotiations with respect to these applications.

In 2014, the Queensland Parliament passed the Mineral and Energy Resources (Common Provisions) Act 2014, which came into effect on 27 September 2016. Under the new legislation, coal is generally granted the right of way over CSG activities, with the coal party being granted sole occupancy rights over discrete areas and the CSG party retaining the right to develop outside the coal areas.

5.4 Domestic gas reservation

Australia is heavily reliant on coal-fired electricity generation; however, there has been a renewed emphasis on moving to lower-emission forms of energy, including natural gas. Given that higher prices for gas can be obtained by producers in the Asian LNG market, there is some concern that gas producers will prefer to divert gas to higher-priced LNG export markets in preference to renewing long-term lower-priced domestic supply contracts. This may result in higher prices being paid in the Australian domestic market, which may impact future economic growth and the economics of switching to natural gas as a transition fuel to a low-carbon economy.

In response to these concerns, there have been calls for states to impose mandatory domestic gas reservations, where producers are to set aside a nominated percentage of gas produced for export for supply into the domestic market.

At this time, Western Australia is the only state which has imposed a domestic gas reservation, requiring that producers commit to supplying a quantity of domestic gas equal to 15% of LNG production from a particular project.

The Queensland Government has the ability to reserve fields for domestic gas production but has not, to date, used this power to reserve any gas for the domestic market. In a discussion paper released in November 2016, the Queensland Department of Natural Resources and Mines also recommended against the use of the reservation policy.[89]

The Australian Federal Government does not support calls for gas export limits and supports free trade of energy.[90] Instead, it has signalled a commitment to

88 *Ibid.*

89 "Queensland gas supply and demand action plan", Department of Natural Resources and Mines, (November 2016) https://www.dnrm.qld.gov.au/__data/assets/pdf_file/0007/805552/gas-action-plan-5107-discussion-paper.pdf.

90 "Election 2016: Labor Moves for national interest test on domestic gas supply", *Australian Financial Review* (17 May 2016) www.afr.com/news/politics/election-2016-labor-moves-for-national-interest-test-on-domestic-gas-supply-20160517-gox8wx.

91 The Hon Josh Frydenberg MP, "Address to the Australian Petroleum Production and Exploration Association Conference" (6 June 2016) www.joshfrydenberg.com.au/guest/SpeechesDetails.aspx?id=371.

boosting gas exports through cutting red tape and through Australia's free trade agreements with China, Japan and Korea.[91] The Federal Government's view broadly aligns with the view expressed by gas producers.

This issue continues to be a matter of ongoing debate as governments try to balance the need to fulfil domestic gas supply needs with the need to sustain and develop the LNG export industry and the Australian economy as a whole.

5.5 Social licence to operate

The term 'social licence to operate' originated in the mining industry and is now firmly entrenched in the lexicon of the CSG industry. The social licence is the level of acceptance or approval continually granted to an organisation's operations or project by the local community and other stakeholders. It is underpinned by the long-term commercial value that companies attach to reputation and trust with all stakeholders in a company's business.

The level of acceptance or approval of key stakeholders such as agricultural groups and local communities towards an organisation and its projects is a key determinant in establishing whether a company has the social licence to operate in respect of its CSG activities.

At the heart of achieving and building social licence to operate is the ability of a company to deal with the issues addressed in this chapter, but also extends to:

- environmental performance, including air and water pollution, water extraction, impacts on biodiversity and waste management;
- ethical business conduct and transparency; and
- safety.

For several of the CSG project proponents, a social licence to operate is a key plank of exploration, development and production activities, and companies are mindful that a social licence to operate, while difficult to obtain and maintain, can be easily lost.

6. The future of unconventional gas in Australia

Australia's unconventional gas industries have undergone significant changes in the last decade. CSG has quickly grown to become an established and important factor in Australia's gas and energy industries, both domestically and internationally. This development has been driven largely by the demands of three Queensland LNG export projects.

While Australia has significant shale and tight gas resources, these resources remain largely undeveloped. Large-scale exploration of these resources is unlikely to occur while the petroleum majors continue to grapple with the impact of the fall in the global oil price. The sharp price fall caused many petroleum companies (particularly those with exposure to oil projects or LNG projects with oil price-linked contracts) to restructure their exploration and development priorities in an effort to reduce cost. Early-stage projects such as Australian shale and tight gas exploration projects are the early victims of such restructuring.

The unconventional industry faces other challenges, not encountered to the same degree as the conventional oil and gas industry. Stakeholder management, social licences to operate and the changing regulatory environment as the government grapples with the issues presented by unconventional gas projects all present significant challenges to these projects. Recently announced moratoria and other legislative measures targeting the CSG industry signal an increase in political uncertainty for the industry.

Despite these challenges, unconventional gas exploration is important for the longer-term energy supply in Australia. The Queensland LNG projects continue to offer Australian gas producers access to energy-hungry markets in the Asia Pacific region. This, coupled with the inevitable field depletion of conventional gas sources such as the Cooper Basin, could lead to gas supply shortage in the east coast from 2020.[92] This will mean that the east coast of Australia has to develop alternative means of securing its gas supply. The recent power blackouts in South Australia (where wind and solar power account for 40% of the state's electricity production[93]) suggest that gas should continue to be a base load energy source to assist with the transition to greener energy sources. It remains to be seen whether these demand-side pressures will encourage continuing development of the unconventional gas industry. One certainly hopes so.

The authors appreciate the contribution of Brodie Same (Senior Associate, Herbert Smith Freehills) and David Douglas (Solicitor, Herbert Smith Freehills) in the preparation of this chapter.

---

92 "Upstream Petroleum and Resources Working Group Report to COAG Energy Council on Unconventional Reserves, Resources, Production, Forecasts and Drilling Rates", Geoscience Australia, (November 2014) www.scer.gov.au/sites/prod.energycouncil/files/publications/documents/Unconventional%20Reserves%20Resources%20Production%20Forecasts%20and%20Drilling%20Rates%202014_0.pdf.

93 "Clean Energy Australia Report" (2015), Clean Energy Council, https://www.cleanenergycouncil.org.au/policy-advocacy/reports/clean-energy-australia-report.html.

Shale gas in the United Kingdom and Europe

Joseph Dutton
Energy Policy Group, University of Exeter

1. Introduction

The first version of this book, published in 2012, set out the challenges that face Europe and other regions in replicating the tremendous success of the shale gas sector in the United States. A range of issues faced shale developers and the nascent sector in Europe was presented with a series of stumbling blocks: some specific to shale gas, some applicable to energy markets in general, and others representative of the problems facing unproven or unfamiliar – in local markets, that is – technology. These extended across all sectors, and the book detailed financial and market place challenges, upstream and technical challenges, environmental challenges, and social challenges.

Four years later and some of these challenges remain. The machinations and evolution of the global gas market has continued – notably with the prospect of the United States exporting shale as liquefied natural gas (LNG) – and profitability remains a challenge. Oil and gas prices have fallen considerably, with little sign of recovering to the highs of earlier this decade. The environmental and climate change case against fracking and shale gas development has been maintained, despite natural gas contributing to the rapid decline in coal-fired power generation. The nature of the sector means these fears have not subsided; rather, in the United Kingdom, as it appears development has come closer, so opposition has become more vociferous. And with this, social opposition to development is greater than before and it may well be the largest barrier to development. Protests on the ground, court cases and government intervention have all ensured fracking has remained in the media spotlight, and obtaining a 'social licence' to operate remains a key issue for the sector.

Although the overall sector has moved on very little since 2012, in the United Kingdom there is greater clarity now on the government's position on shale development. Elsewhere in Europe, the intervening years have also given a clearer picture of where development of shale is now unlikely to happen. This chapter will provide background to UK government policy and the policy, planning and development frameworks for the shale gas sector; furthermore, it will update the status of development in other European countries that had previously been regarded as future shale producers. It will also set out how the development of the sector could be affected by the United Kingdom leaving the European Union.

2. US shale: unconventional becomes the convention

In the United States, the shale gas 'revolution' has firmly matured and made the progression to an established sub-sector of oil and gas. Overall natural gas output has continued to grow through this decade off the back of the growth in shale, and in 2015 production reached its highest level since records began in 1930. Shale gas reached a 40% share of total gas output in 2013 from only 8% in 2007, and that year production from shale gas wells overtook that from other wells for the first time, reaching 33 billion cubic feet per day (bcf/d) compared to only 5bcf/d in 2007.[1] The ubiquitous nature of shale gas and fracked wells means the use of the term 'unconventional' to describe shale gas and its production methods has been rendered inaccurate. In 2000 the United States had 26,000 active natural gas wells that had undergone fracking; by 2015 this total had reached over 300,000, with fracked wells making up 67% of total gas production.[2] A similar trend can be seen in the oil sector, where fracked wells in 2015 accounted for 51% of output compared to just 2% in 2000.

When the first edition of this book was written in 2012, discussions were emerging about the United States exporting natural gas as LNG to European and Asian markets. That year the United States had just one LNG export terminal, while a further three re-export terminals had been approved and seven formal terminal proposals had been received by the Federal Energy Regulatory Commission (either as new-build or expansion of existing import terminals). By 2015 overall US exports of gas reached their highest level on record, with net imports dropping to their lowest level since 1986. Much of this growth was because of higher deliveries to Canada and Mexico – the two largest recipients of US gas – but LNG exports are also increasing. And highlighting the very different stages of 'revolution' that the US and UK shale gas sectors find themselves at, in September 2016 a cargo of ethane – a by-product of natural gas, used in the chemical industry – arrived at the Grangemouth refinery from the United States. Chemical company Ineos signed a 15-year contract with a US oil company for ethane, and intends to import up to 800,000 tonnes per year from the United States to its refineries in Grangemouth and Rafnes, in Norway.[3]

3. Status of the UK shale sector

In May 2016 Third Energy was given planning permission by North Yorkshire County Council to undertake fracking operations and test drilling for shale gas near the small village of Kirby Misperton.[4] This was the first new project in the United Kingdom to be approved since 2011. The early part of this decade had been marked by an extreme optimism on the part of those involved in the development of shale gas, but development has yet to become a reality. At the time of writing the previous edition of this book in 2012, the sector had not come close to anything resembling commercialisation; and today, moving towards the end of the decade, little has changed.

---

1 www.eia.gov/todayinenergy/detail.php?id=18951.

2 US Energy Information Administration, 5 May 2016, "Hydraulically fractured wells provide two-thirds of US natural gas production".

3 INEOS/ICIS, April 2016, "INEOS rides the shale wave".

4 North Yorkshire County Council, 23 May 2016, "Fracking decision".

Between 2011 and 2012 the sector was at a standstill because of the government-imposed moratorium on fracking. In Spring 2011 fracking operations at Cuadrilla's Preese Hall site in Lancashire were highly likely to have caused seismic activity, although no damage was reported at the surface. Following an investigation by the government and Cuadrilla, all fracking operations were halted under a moratorium that lasted from November 2011 until December 2012.

Fracking firmly became a headline topic in 2013, due in part to the government reaffirming its desire for development, but also because of large-scale protests at Cuadrilla's operation at Balcombe in Surrey. Crowds of campaigners and protesters, accompanied by a large police presence, camped in the village through the summer and ensured widespread media coverage (see section on social challenges). Yet despite the protests and largely negative media coverage, heading in to 2014 prospects seemed more positive. In June 2013 it was announced that shale resources in northwest England were almost double the previous estimate.[5] International chemical firm Ineos also announced its intention to invest around £700 million to develop its shale gas licences in Scotland and establish supply of gas locally for its Grangemouth petrochemical plant. However, as detailed above, Ineos later moved to import ethane from the United States, such has been the slow pace of development in the United Kingdom.

In early 2015 the pendulum swung back and the sector received a setback when Lancashire County Council recommended that an application from Cuadrilla to drill four wells at its Preston New Road site should be rejected – even though the Environment Agency had already granted a permit to drill at the site. The permit had set out conditions regarding ground and surface water, air quality and wastewater disposal, and the regulator was confident that the right controls were in place. However, the local authority decided that not enough had been done to tackle noise pollution and disturbance to local residents. An application to drill four wells at the nearby Roseacre Wood site was also rejected due to noise and traffic concerns, but the planning authorities rejected concerns about environmental pollution and water contamination.

Although Cuadrilla appealed the verdict and the government overturned the local authority's decision (see next section on government policy), the UK sector still remains far from commercialisation. Of the 11 wells that were due to be drilled in 2015, eight were halted following Lancashire planning authorities' decision on the two Cuadrilla sites. But the shale sector's inactivity is reflected across the wider onshore oil and gas sector. Only 14 exploration wells were drilled in 2013–2015 compared to 18 wells in 2011–2012; and only one well was drilled in the first half of 2016.[6] Onshore operations for oil and gas now face increased levels of scrutiny that they did not before, even in regions where conventional production is already established.

5 *BBC News Online*, 27 June 2013, "UK shale gas resources 'greater than thought'".

6 Department for Business, Energy and Industrial Strategy, 25 November 2016, "Energy Trends: Drilling on the UKCS".

4. Government policy and the political climate

Since the first shale gas exploration well was drilled in Lancashire in 2011 the United Kingdom has gone through a tumultuous political period. There have been three different phases of government and two prime ministers: the Conservative-Liberal Democrat coalition (2010–2015), David Cameron's Conservative government (2015–2016), and now fellow Conservative Theresa May's government. Despite the different government phases, developing shale has remained a priority.

4.1 2010–2015: Conservative-Liberal Democrat coalition

Fracking was a divisive issue under the coalition, particularly as the position of secretary of state for energy and climate change and head of the Department for Energy and Climate Change was held by Liberal Democrat MPs. As a party, the Liberal Democrats have consistently been opposed to fracking, but their Conservative partners in the coalition were very much in favour of development. Liberal Democrat leader and then Deputy Prime Minister Nick Clegg even spoke out publicly against fracking and shale gas development because of extensive environmental damage caused by the sector in the United States.[7]

In December 2012 the 13 month-long fracking moratorium that was introduced following seismic activity at Cuadrilla's Preese Hall site in 2011 came to an end, and Energy Secretary Ed Davey announced development could proceed, with new controls on seismic activity detection and prevention added to existing regulations.[8] As part of the new controls the government also created the Office of Unconventional Gas and Oil within the Department of Energy and Climate Change. The office was tasked with a number of roles, including making the most of UK natural resources, enabling energy development while ensuring public and environmental safeguarding, and ensuring local communities benefit from development.[9] David Cameron publicly stated his intention to go "all out for shale"[10] and it was during this period that government policy on shale began to take shape, including changes to the tax regime to encourage shale development, a government-backed exploration fund of £5 million to demonstrate to the public the robustness of planning regulations, new planning guidance from the Department for Communities and Local Government, and new permitting rules for the Environment Agency to speed up development licence applications.[11] The large-scale protests at Balcombe (see section on social challenges) in the summer of 2013 provided unwanted negative coverage for the shale sector, although the government was undeterred in pushing for development. Towards the end of the year, Energy Minister Michael Fallon restated the government's desire to develop shale, and set out a regulatory roadmap detailing the process operators should follow when seeking

---

7 *The Guardian*, 24 August 2013, "Liberal Democrats blast environmental damage cause by fracking".

8 Department of Energy and Climate Change, 13 December 2012, "New controls announced for shale gas exploration".

9 Department of Energy and Climate Change, "Office of Unconventional Gas and Oil" https://www.gov.uk/government/groups/office-of-unconventional-gas-and-oil-ougo.

10 *The Guardian*, 13 January 2013, "Fracking in the UK: we are going all out for fracking, admits Cameron".

11 Department of Energy and Climate Change, 11 April 2016, "Guidance on fracking: developing shale oil and gas in the UK".

to drill for onshore oil and gas – whether fracking is involved or not.[12] The Department of Energy and Climate Change also released an explainer entitled 'Developing Onshore Shale Gas and Oil – Facts about Fracking', which argued strongly in favour of its development.

Following on from the roadmap, in its 2014 election manifesto the Conservative party said it was determined to ensure local communities benefitted from any development, but gave little detail on what this would look like. Although this was the first time it had entered the policy arena in earnest, the proposal built on the UK Onshore Operators Group voluntary Community Charter of 2013. Commitments in the charter included providing benefits to local communities at the exploration and appraisal stage of £100,000 per well site where fracking takes place, with a share of proceeds at production stage of 1% of revenues allocated on an approximate basis of 66% to the local community and 33% at the county level.[13]

In January 2015 the Environmental Audit Committee called for a second moratorium on fracking in the United Kingdom – this time because of shale's incompatibility with tackling climate change and the fact that developing the sector would extend the country's dependence on fossil fuels.[14] However, the government countered that shale is compatible with goals to cut greenhouse emissions as natural gas reduces the need for coal in power generation, and that shale would not detract from government support for renewables because gas-fired power generation is needed to support development of intermittent renewables.[15] The proposed moratorium was overwhelmingly rejected by 308 votes to 52 by MPs, although the opposition Labour party did secure 13 amendments to the government's related Infrastructure Bill that ensured tighter regulation for fracking. The bill set out conditions and safeguards, and required documentation from regulators for fracking to take place.[16] Although a number of these duplicated existing onshore hydrocarbon legislation, it was the first time they were set out explicitly in reference to shale gas and fracking. Key elements of the Infrastructure Bill included the prohibition of fracking at a depth shallower than 1,000 metres – although shale formations are typically found much deeper – and the so-called underground access clause, which simplified the procedure for accessing land to depths of more than 300 metres for the oil and gas extraction and geothermal energy.[17]

4.2 2015–2016: Conservative (David Cameron)

Ahead of the 2015 general election, energy had been expected to play a substantial role in voting. The so-called 'cost of living crisis' and muted wage growth following the recession ensured that consumer bills and energy costs became a headline issue.

---

12 Department of Energy and Climate Change, 17 December 2013, "Publication of a regulatory roadmap for onshore oil and gas exploration in the UK and a strategic environmental assessment".

13 UK Onshore Operators Group, 2013, "Community Engagement Charter Oil and Gas from Unconventional Reservoirs".

14 www.publications.parliament.uk/pa/cm201415/cmselect/cmenvaud/856/85607.htm.

15 Department of Energy and Climate Change, 26 March 2015, "Government response to the environmental audit committee's inquiry into environmental risks of fracking".

16 HM Government, Infrastructure Act 2015 – Part 6, energy.

17 Department for Business, Energy and Industrial Strategy, December 2016, "National Preventive Action Plan: Gas".

Labour leader Ed Miliband announced he would freeze energy prices if elected.[18] But even though it received widespread media coverage, it made very little difference to voter intention. And despite the Environmental Audit Committee call for a new fracking moratorium and the concessions Labour gained from the government, it too made little difference in the election. Fracking had previously been identified as a key topic by some; Greenpeace even launched a 'Frack Free Promise' pledge, which 54 MPs signed.[19] Rather tellingly perhaps, only two signatories were from the victorious Conservative party, and the overwhelming majority were from constituencies where there are no shale gas resources.

Following election victory in May 2015 – and the end of the coalition – the Conservative government reinforced its intention to develop shale gas. Within three months of the election, the Department of Energy and Climate Change and the Department for Communities and Local Government released a policy paper that made the case for developing shale gas and oil.[20] The paper primarily argued for development because of energy security concerns on declining North Sea gas production, and the economic and employment benefits the sector could deliver – the two main lines of justification that have regularly been used for developing shale. This was soon followed by a statement from minister of state Andrea Leadsom sub-entitled "an inconvenient truth for the anti-fracking lobby", which strongly argued for developing shale in the context of "keeping the lights on, keeping the bills down, and moving to a clean energy future".[21]

4.3 2016–present: Conservative (Theresa May)

Following the EU membership referendum result and the subsequent resignation of David Cameron in June 2016, under new Prime Minister Theresa May the Department for Energy and Climate Change was disbanded and replaced by the Department for Business, Energy and Industrial Strategy in July 2016. The government signalled its intent to link energy and industrial policies, and the new department maintains the energy and climate-related policies of the Department for Energy and Climate Change.

Although the focus for this government will undoubtedly be the country's exit from the European Union, it signalled its intent to push for shale gas development and furthered previous suggestions to make payments to affected communities. It proposed using a new Shale Wealth Fund to distribute 10% of all shale gas tax revenues to local communities, up to a maximum pay out of £10 million to each community associated with a shale drilling site over a 25-year period. The government also proposed that payments could be made directly to households, rather than only as benefits to communities or local authorities.[22] But later in the

18 *BBC News Online*, 24 September 2015, "Ed Miliband: Labour would freeze energy prices".
19 http://greenpeace.org.uk/blog/climate/which-mps-have-spoken-out-against-fracking-here-are-few-keep-eye-20150615.
20 Department of Energy and Climate Change, 13 August 2015, "Shale gas and oil policy statement: DECC and DCLG".
21 Department of Energy and Climate Change, 23 September 2015, "Shale gas – an inconvenient truth for the anti-fracking lobby".
22 HM Treasury, August 2016, "Shale Wealth Fund: consultation".

year, the government made arguably the most defining action in shale policy. In June 2015 Lancashire County Council refused planning permission for Cuadrilla to carry out exploration activity and fracking at its Preston New Road and Roseacre Wood sites. Cuadrilla appealed and the government overturned the local planning authority's decision on Preston New Road in October 2016.[23] The secretary of state for communities and local government has the power to take over ('call in') planning applications and override local authorities. This is normally only done if the application conflicts with national policy or it is of national significance.

The decision to override the local authority signified the government's intent to develop the shale sector, as part of its new industrial strategy in the newly created Department for Business, Energy and Industrial Strategy. But the decision was ultimately extremely divisive, with local communities reacting very strongly against the decision amid accusations of being ignored by government and undermining democracy.[24] Importantly for the sector, this will be the first site to use a combination of fracking with horizontal drilling, rather than solely vertical drilling as has been done at other developments. The opposition Labour party announced in September 2016 that it would ban fracking if elected at the next general election, primarily because of it being incompatible with tackling climate change.[25]

5. Social challenges

Understanding the geological and financial characteristics of developing shale gas in the United Kingdom remains a work in progress and could prove challenging in the medium to longer term; however, social challenges and public opinion remain the sector's biggest inhibiter both now and going forward. The government's intervention at the Preston New Road site has worried other communities, and highlighted the tension between local impacts of development and broader considerations of energy security and economic activity. However, in general, there is a lack of a 'social licence' to operate, even if companies have been given the correct permits that legally allow them to operate.

Public awareness on fracking and shale gas outside of the United States developed before operations began in the United Kingdom, primarily due to the use of social media by environmental groups and opponents. The 2010 Oscar-nominated but widely challenged anti-fracking documentary Gasland – which included an infamous clip of water from a tap being ignited due to high levels of methane, allegedly because of fracking – received coverage across the world, and its director Josh Fox remains an active campaigner both in the United States and Europe. Media coverage of shale gas and fracking in the United Kingdom remained within more specialist financial and industry-related outlets, but it became firmly mainstream in the spring of 2011 following the suspension of fracking operations by Cuadrilla at its Preese Hall well site in Lancashire because of seismic activity. Such was the strength of opposition to fracking, and the highly organised nature of environmental groups'

---

23 Lancashire County Council, "Shale gas developments in Lancashire" www.lancashire.gov.uk/council/planning/major-planning-applications/shale-gas-developments-in-lancashire.aspx.

24 *BBC News Online*, 6 October 2016, "Fracking in Lancashire given go-ahead by government".

25 *BBC News Online*, 26 September 2016, "Labour vows fracking ban if it wins general election".

social media campaigns, that even after the moratorium on fracking was lifted and new safeguards were introduced, opposition remains high. Conventional oil and gas operations also began to come under scrutiny – even in areas where oil and gas production had been occurring for decades.

The most notable example of this was in Balcombe, in southern England. Site developer Cuadrilla had been given planning permission by local authorities to undertake exploratory drilling operations in a conventional oil well – not shale gas – at Lower Stumble, which was originally drilled in 1986 by Conoco. There was no fracking in the licence application or planning permission, but in mid-August 2013 the site became a focal point for anti-fracking operations and a 'climate camp' was set up in a nearby field under the direction of the environmental action group 'Reclaim the Power'.[26] The camp attracted protestors from a range of organisations, including Greenpeace, Friends of the Earth, WWF and InfoWars, and people without affiliation also travelled to the village. The proximity to London and good train connections meant people continued to arrive at the camp daily, including public figures such as fashion designer Vivienne Westwood and Green Party MP Caroline Lucas, along with television, radio and print media, who reported from the site daily. The protests at the village lasted for 68 days and a total of 124 people were arrested, according to campaigners.[27]

Misinformation on the nature of developments can be observed elsewhere in the United Kingdom too. For example, developments at Kiln Lane in Lincolnshire and Horsehead Hill in West Sussex appear on the website of campaign group FrackOff as shale gas developments, but both of them are oil and at neither site were planning and licensing applications made for fracking. Similarly, the website lists at least four other sites in southern England as being for shale gas that are in fact prospective oil developments. And some of these are in locations where oil production is already established, such as near to Wytch Farm in Dorset – the largest onshore oil field in Western Europe. In the Sussex Downs area of southern England, where the Horsehead Hill and Lower Stumble developments are found, there are already around 50 oil and gas wells; however, localised campaign and environmental groups have emerged only in recent years.

Irrespective of the accuracies of some campaign organisations' reporting, since 2011 a groundswell of localised and national opposition groups has developed. Campaign groups can be found across regions where both shale gas and other onshore oil and gas developments are ongoing or planned. And with a significant social media presence, larger regional and national groups have been successful in orchestrating campaigns. The example above of Balcombe in 2013 serves as a primary example, but there are others; for example, North Yorkshire County Council received more than 4,300 objections to prospective development and fracking at the Kirby Misperton site, even though the village has a population of only 370 people.[28]

Although the current government has gone further than previous ones in placing

---

26 https://reclaimthepower.org.uk/2013campbalcombe/

27 http://frack-off.org.uk/fracking-balcombe-a-threatened-community-2010-2013/

28 Office for National Statistics, "Neighbourhood Statistics – 2011 Census".

communities at the centre of development with the suggestion they could receive a share of profits from shale gas operations, public support for fracking remains low. In surveys taken by the Department for Energy and Climate Change in the spring of 2016 – before the new fund and pay out proposals – public support for fracking stood at just 19%, while 31% were explicitly opposed to it.[29] The government's proposal to distribute payments to the communities and households was arguably an attempt to win over the public and speed up the development process, but they were met with some resistance. A Green Party MEP described them as "immoral and tantamount to bribery" – a view also echoed by other MPs.[30] In the wake of the proposals a poll commissioned by Friends of the Earth found that 33% of people would support fracking if households were paid, but 43% would still oppose fracking regardless of payments.[31]

6. Shale gas and Europe

The focus of this chapter has so far been on the United Kingdom, but other European countries are also estimated to have substantially higher shale gas resources. The US Energy Information Administration carried out resource assessments for selected European countries in 2011, and again in 2013 and 2015. Yet although resources are higher, development across Europe has been as slow – or even slower – than in the United Kingdom, and some places have introduced all-out bans.

Table 1: Selected European unproved technically recoverable shale resources (tcf)[32]

Bulgaria	17	Romania	51
Denmark	32	Russia	285
France	137	Spain	8
Germany	17	Sweden	10
Netherlands	26	Ukraine	128
Poland	146	United Kingdom	26

6.1 Poland

Poland was at one time considered the front-runner of European shale gas development, even ahead of the United Kingdom by some. Suggested benefits of

29 Department of Energy and Climate Change, April 2016, "Public Attitudes Tracker – Wave 17".
30 *The Guardian*, 7 August 2016, "Trying to bribe the public to accept fracking won't work, say campaigners".
31 https://drillordrop.com/2016/08/15/pms-fracking-payment-plan-fails-to-win-majority-support-for-shale-gas-poll/.
32 US Energy Information Administration, 2015, "World Shale Gas Resource Estimates" https://www.eia.gov/analysis/studies/worldshalegas/.

development were similar to the United Kingdom in terms of economic and employment growth, as well as the environmental effect of replacing coal with gas in electricity generation; but the energy security aspect was the main driver – specifically reducing the country's dependence on Russian gas.[33] Despite the optimism as far back as 2012, some problems began to emerge. Two wells drilled by US company ExxonMobil did not have economic flow rates, while those drilled by 3Legs Resources and BNK Petroleum had flow rates that were much lower than expected. That year drilling activity reached its peak, with 24 wells drilled, and exploration licences covered nearly 30% of the country. But activity fell away, and Poland has not drilled a shale gas well since 2015. The total number of exploration licences held by companies plummeted to 31 at the start of 2016 from a high of 115 only three years earlier. After international companies exited Poland, the last two active developers, domestic leaders PKN Orlen and PGNiG, stopped their shale gas exploration activity in October 2016.[34]

Table 2: Poland shale drilling and fracking activity 2010–2016[35]

Year	Wells drilled	Fracked wells
2010	3	2
2011	12	7
2012	24	8
2013	14	4
2014	15	4
2015	4	0

6.2 Denmark

In 2013 Denmark was estimated to have resources of up to 32tcf – higher than the United Kingdom's 26tcf – with the majority in the Alum Shale formation, which also extends into Sweden.[36] In 2012 the government imposed a moratorium on new licences for exploration and development of shale gas, but this did not apply to those issued before 2012. French company Total drilled exploration wells in both of its licence areas, but decided to exit Denmark in 2016 because of results. Shell drilled

---

33 C Johnson & T Boersma, "Energy (in)security in Poland: the case of shale gas" *Energy Policy*, 53, (2013) 389-399.

34 *Reuters*, 12 October 2016, "Polish firms concede defeat in search for shale gas riches".

35 Polish Geological Survey, 11 April 2016, "Shale gas exploration status in Poland as of April 2016".

36 US Energy Information Administration, 2015, Technically Recoverable Shale Oil and Shale Gas Resources: Other Western Europe.

three wells for gas in the Swedish section of the Alum between 2008 and 2011 but also reported uneconomic flow rates.

6.3 Germany

Energy security was a major factor in calls for Germany to develop its shale gas resources, as domestic gas production accounts for less than 10% of total consumption. There was also an environmental dimension to this, through using natural gas to replace coal in power generation alongside renewables. Fracking and water-based stimulation in conventional and tight gas production dates back to the 1950s, but it was not done specifically for shale gas until 2008.[37] In 2015 Germany initially came close to passing a ban on fracking, but political disagreements between the Grand Coalition government parties prevented the legislation. A year later though, in June 2016, the parties did agree a ban on fracking, with the condition it would be revisited by Parliament in 2021.[38]

6.4 France

France was estimated to have the highest shale gas resources in central and western Europe, but it too has passed a ban on fracking. A total of 64 exploratory permits were issued to 20 companies in 2010, but following a groundswell of strong opposition in 2011 a six-month moratorium was imposed. In July that year Parliament voted to ban fracking outright and three permits that involved fracking were revoked. The other 61 remained active as the companies demonstrated they were not using fracking.[39] Two years later, in 2013, the ban was challenged at the constitutional court by US company Schuepbach Energy, which lost its exploration permits when the ban was introduced; however the ban was upheld.[40]

7. Shale gas and Brexit

The UK shale gas industry and, more generally, energy and natural gas sectors, are not immune from the potential upheaval that leaving the European Union will have. Energy was not a headline issue during the referendum campaign and it is unlikely to feature during the coming negotiations; however, the key issues of access to the single market (both in general and, specifically, the internal energy market) and the future of current EU legislation in the United Kingdom are of direct relevance to shale gas development.

The relationship that the United Kingdom will seek to negotiate with the European Union for when it leaves has not yet been publicly announced, despite the government suggesting it will seek a so-called 'hard Brexit'. But there are various existing options the United Kingdom could pursue for its future relationship with the European Union. The cases of Norway and Switzerland are useful but not definitive examples of third-party country relations with the European Union;

---

37 Shale Gas Information Platform, August 2016, "Shale gas in Germany – the current status".
38 *Reuters*, 21 June 2016, "German government agrees to ban fracking indefinitely".
39 R Weile, "Beyond the fracking ban in France", *Journal of European Management and Public Affairs*, 1 (2), (2014).
40 *Bloomberg*, 11 October 2013, "France's fracking ban 'absolute' after court upholds law".

indeed, the United Kingdom could press for a more bespoke relationship than those currently enjoyed by Norway and Switzerland. The eventual nature of UK–EU relations will have a strong bearing on the future of energy in the United Kingdom as the two parties are strongly linked in energy supply and policy terms.

7.1 Single market access

The future economic relationship between the United Kingdom and the European Union is likely to be a central point of negotiation. Membership of the single market or access to it with equivalence to the United Kingdom's current level is something that business leaders and politicians have called for. The government has given little or no detail on what trade regime it will seek after Brexit, but the European Commission appears to be steadfast in its position that membership of the single market requires freedom of movement of people – something pro-Brexit politicians said would end after the United Kingdom leaves the European Union. In energy terms, the internal energy market is crucial for the UK energy sector, and companies across the energy sector have called for the United Kingdom to remain part of it; however, access to it could affect the development of shale gas in negative and positive ways.

Wider economic consequences of leaving the European Union would have a secondary effect on development, as has already happened with the wider energy and industrial sectors. Reduced economic activity lowers industrial output, in turn reducing energy demand and consumption. A worsening exchange rate also increases the cost of importing equipment for the development supply chain. Developer Cuadrilla imported its drilling and fracking units from Italian company Drillmec for its operations, and leaving the single market could also lead to increased costs through the imposition of tariffs. The oil and gas industry has also expressed its concern about the recruitment of skilled labour in the event of leaving the single market.[41]

If the United Kingdom leaves the single market and tariffs are imposed on UK gas exports to the continent, via the IUK and BBL pipelines, this could result in reduced gas supply for the UK market; in effect, if the United Kingdom becomes less connected with the European marketplaces, supply security could decrease. Typically, in winter months the United Kingdom draws heavily on gas storage facilities on the continent. Therefore, the need for developing indigenous resources becomes more pressing if this supply falls, and this could spur the government to take further action to ensure development. Conversely, if exports decrease through becoming more disconnected (in a marketplace sense, rather than physically), this could increase available supply for the UK market and reduce the need for developing more gas. Security of supply has been a primary point of justification for developing shale, but this could weaken the case for its development. The premise for this idea can be seen in the electricity sector, where there is substantial trade between the United Kingdom and Europe. In November 2016 the IFA interconnector between the United Kingdom and France was damaged, leading to a reduction in its

41 *Financial Times*, 24 October 2016, "Brexit and the energy sector – the risks of Little Britain".

capacity. Although this was widely reported as a threat to security of supply for the United Kingdom, the IFA typically exports to France in winter; therefore, it being damaged meant reduced electricity exports and more supply for the United Kingdom domestically.[42]

7.2 European legislation

Energy policy is broadly a member state competence in the European Union, and the right of member states to determine the make-up of their energy mix is enshrined in the Lisbon Treaty. However, secondary environmentally focused policies have altered the shape of energy and electricity generation in member states; for example, directives promoting renewable energy development and those aimed at reducing pollutants and emissions which have led to the closure of coal-fired power stations. In the context of shale gas, a number of key EU directives and legislative frameworks are in force in the United Kingdom that regulate the development of the sector, and also feed into and underpin associated UK-only policies (see Table 3 on the next page). These are focused predominantly on environmental quality and protection, and sit alongside energy and hydrocarbon-specific regulation developed by the United Kingdom.

The future presence and legal effect of these directives in the United Kingdom is unknown, as the government has not yet disclosed what political and legal relationship it will seek to have with the European Union when the United Kingdom leaves. The current timetable for exiting is Spring 2019, but even this is subject to revision as the government could delay triggering 'Article 50' and the start of the formal two-year exiting process. In October 2016 the government said that existing EU legislation in force in the United Kingdom would be written across into UK law upon leaving the European Union, with the creation of the Great Repeal Bill. This bill would repeal the European Communities Act 1972 – which provides legal authority for EU law to have effect as national law in the United Kingdom – and subsequently transfer all EU laws currently in force onto the UK statute book.[43] Even so, it remains unclear exactly which bits the government would want to keep or discard, and whether they would be changed or updated in line with changes made in the European Union in the future.

---

42 Argus Media, 30 November 2016, "UK power market shrugs off IFA limitation".

43 Institute for Government, "Brexit Brief: the Great Repeal Bill" https://www.instituteforgovernment.org.uk/brexit/brexit-brief-great-repeal-bill.

Table 3: Key EU Environmental Legislation in England, Wales and Scotland[44]

EC Legislation	Associated UK Legislation	Main Requirements	Regulator, Applies in
EC Directive (85/337/EEC): Assessment of the effects of certain public and private projects on the environment	1. Town & Country Planning (Environmental Impact Assessment) (England and Wales) Regulations 1999 2. Environmental Impact Assessment (Scotland) Regulations 1999	Requires certain developments to prepare an environmental statement as part of the planning approval process	1. Local authorities, England and Wales 2. Local authorities, Scotland
EC Directive (92/43/EEC): Conservation of natural habitats and of wild fauna and flora	Conservation (Natural Habitats) Regulations 1994	Requires developments to take account of special areas of conservation in their environmental impact assessment. Approvals granted via the above regulations	Environment Agency or English Nature, England and Wales Scottish Environment Protection Agency or Scottish Natural Heritage, Scotland

continued on next page

44 Department for Business, Energy and Industrial Strategy, "Environmental legislation applicable to the onshore hydrocarbon industry" https://www.gov.uk/guidance/oil-and-gas-onshore-exploration-and-production.

EC Legislation	Associated UK Legislation	Main Requirements	Regulator, Applies in
EC Directive (96/82/EC): Control of major accident hazards	1. Control of Major Accident Hazards (COMAH) Regulations 1999 2. Planning (Control of Major Accident Hazards) Regulations 1999 [2000 in Scotland]	Authorisation is required for storage of listed hazardous substances Requires operators to implement certain management practices and report to the competent authorities	1. Environment Agency and local authorities, England and Wales 2. Scottish Environment Protection Agency and local authorities, Scotland
EC Directive (80/68/EEC) old Groundwater Directive (in force till Dec 2013); and (2006/118/EC) Groundwater Daughter Directive EC Directives 2006/118/EC and 2008/105/EC	1. The Environmental Permitting Regulations in England & Wales 2. The Water Environment (Controlled Activities) (Scotland) Regulations 2011	Systems of permits and registrations to control inputs of pollutants to the water environment	1. Environment Agency, England and Wales 2. Scottish Environment Protection Agency, Scotland

continued on next page

EC Legislation	Associated UK Legislation	Main Requirements	Regulator, Applies in
Water Framework Directive	1. The Environmental Permitting Regulations 2. The Water Environment (Controlled Activities) (Scotland) Regulations 2011	Prevent deterioration and achieve good status for all water bodies, reduce pollution from priority substances in surface waters, reverse significant and sustained upward trends in concentrations of pollutants in groundwater, prevent or limit inputs of pollutants to groundwater	1. Environment Agency, England and Wales 2. Scottish Environment Protection Agency, Scotland
Directive 2004/35/EC on environmental liability with regard to the prevention and remedying of environmental damage	1. The Environmental Liability (Scotland) Regulations 2009 2. The Environmental Damage (Prevention and Remediation) Regulations 2009	To introduce a system of reporting and management of significant releases of pollutants to land and the water environment	Scottish Environment Protection Agency, England and Wales
EC Regulation (259/93): Supervision and control of shipments of waste within, into and out of the European Community	Transfrontier Shipment of Waste Regulations 1994	A licence is required to control the transport and disposal of movement and disposal of hazardous waste	Environment Agency, England Scottish Environment Protection Agency, Scotland

continued on next page

EC Legislation	Associated UK Legislation	Main Requirements	Regulator, Applies in
EC Regulation (3093/94): Substances that deplete the ozone layer	1. Environmental Protection (Controls on Substances that Deplete the Ozone Layer) Regulations 1996 2. Ozone Depleting Substances (Qualifications) Regulations 2006 SI 1510 3. Fluorinated Greenhouse Gases Regulations 2008 (SI No 41)	A licence is required for the production, supply, use, trading and emission of certain "controlled substances" that deplete the ozone layer	Department for Environment, Food and Rural Affairs, England, Wales & Scotland
EC Directive 96/61/EC concerning integrated pollution prevention and control	1. The Environmental Permitting Regulations 2. The Pollution Prevention and Control (Scotland) Regulations 2000 (as amended)	Control of emissions from industrial premises through requirement to apply Best Available Technology and Permitting	1. Environment Agency, England and Wales 2. Scottish Environment Protection Agency, Scotland
Industrial Emissions Directive 2010/75/EU	To be transposed into Scottish Legislation by 2012	Brings together into single text previous directives on: integrated pollution, prevention and control; waste incineration; large combustion plants; solvent emissions; and titanium dioxide	

continued on next page

EC Legislation	Associated UK Legislation	Main Requirements	Regulator, Applies in
CCS Directive 2009/31/EC	Energy Act 2008	Sets out requirements for carbon capture and storage	

Shale gas for LNG

Vivek Bakshi
Dentons Canada LLP

1. Introduction

This chapter covers the impact of shale gas on the liquefied natural gas (LNG) sector and its potential as feedstock for LNG export projects in North America. The specific issues arising from the use of shale gas (as distinct from conventional natural gas) as a feedstock are examined, as well as the impact that this may have on the allocation of project risk in liquefaction projects. Finally, the current status of the LNG export market in North America is reviewed.

When the International Energy Agency declared in mid-2011 that the years leading up to 2035 may be a "golden age for gas",[1] the full impact of the Tohoku earthquake and resulting tsunami on the global nuclear power generation industry was yet to be determined. Further, on 4 November 2016, the United Nations Framework Convention on Climate Change (the Paris Agreement) entered into force. The International Energy Agency predicts that, even as Paris Agreement-signatory countries move away from fossil fuels and towards renewable sources of power generation, the world will see strong growth in natural gas generation in the years leading up to 2040.[2] The prospect of natural gas-fired power as a partial replacement for nuclear power, the need to backstop renewable sources of generation with constantly available gas, and the potential for phase-out of coal generation around the world will all play into the evolution of the energy supply mix and have presented themselves as key drivers for the development of LNG export projects in North America, particularly those on the continent's west coast.

Australia has led the way in the development of LNG export projects involving unconventional natural gas, using coal seam gas or coal bed methane as feedstock, although the imperative for the development of an export industry in North America may well be different. The development of North America's significant shale gas resources has contributed to a fall in natural gas prices at all of the continent's trading hubs and has had a significant effect on its nascent LNG industry, prompting the region – which in the mid-2000s was a vibrant, developing LNG import market – to reinvent itself as a potential player in global LNG supply.

---

1 "Are We Entering a Golden Age of Gas?", International Energy Agency, Special Report World Energy Outlook 2011.

2 International Energy Agency, World Energy Outlook 2016.

2. Impact of shale gas on natural gas markets

While the production of shale gas is subject to numerous challenges, the prospect that indigenous shale gas reserves might be an importing nation's solution to addressing energy security and independence is significant. Depending on the source and reliability of a nation's natural gas imports, shale gas production that displaces such imports may have material geopolitical consequences.[3]

The prospect of energy independence in Poland through the production of shale gas has waned in recent years; while significant challenges to the development of the Polish shale gas industry remain, the geopolitical consequences of being independent of Russian-supplied natural gas are still keenly debated.[4] As uncertainty remains regarding the development of a domestic supply, Poland has turned to the construction of LNG import terminals in order to reduce dependence on current sources, with the first terminal receiving commercial cargoes in 2016.[5]

2.1 Effect of flood of supply into North American pipeline delivered market

In North America,[6] shale gas production has not been viewed in the same way as in Poland, largely due to a lack of reliance on 'foreign'[7] natural gas (but see Section 2.2 below). Shale gas production has offset declining production from conventional natural gas resources (due to basin maturity), but initially coincided with the global financial crisis and falling natural gas demand. While demand has since rebounded in North America, natural gas production (fuelled by an increase in shale gas production)[8] has continued to rise. The impact on natural gas prices has been profound and has been discussed in detail elsewhere in this book.

2.2 North American LNG market pre-shale and post-shale

The sustained low natural gas prices in the United States and Canada have had a significant effect on their respective LNG industries. Apart from Alaska's Kenai LNG, investors in the US and Canadian LNG sector had focused their attentions on the import and regasification of LNG; as a result, there are presently eight LNG regasification terminals in the United States and one in Canada. While a number of these regasification terminals have existed in the United States for a considerable time, investment decisions were taken to build a significant portion of the newer regasification capacity during periods of high domestic demand, declining

---

3 See Application of Sabine Pass Liquefaction LLC and Sabine Pass LNG LP for Authorisation under Section 3 of the Natural Gas Act, 31 January 2011, p31, in which the applicant identifies certain geographical and national security interests that might be advanced by LNG exports from the United States.

4 "Europe's Energy Independence Drive Goes off the Rails", *Foreign Policy*, 5 February 2015.

5 *Ibid.*

6 In this chapter, 'North America' is taken to mean the United States and Canada.

7 There has historically been a fluid natural gas market between the United States and Canada, such that the two are often viewed as a single market. The availability of cheap natural gas from the United States is now resulting in natural gas being imported into Southern Ontario from the United States, rather than such Canadian demand being satisfied by production from the Western Canada Sedimentary Basin (which has been declining consistently since 2005).

8 The US Energy Information Administration, in its Annual Energy Outlook 2017 at p57, projects that shale gas will account for nearly two thirds of total natural gas production in the United States by 2040. This represents an increase from 23% in 2010 as reported in the Annual Energy Outlook 2012 at p93. At the start of this century, annual shale gas production in the United States was under 0.5 trillion cubic feet, out of a total production figure of around 20 trillion cubic feet at that time.

conventional supply and high natural gas prices. In fact, in 2004 the US Energy Information Administration was aware of as many as two dozen proposals to build new regasification terminals; it projected that by 2010 the new terminals would be collectively importing into North America 812 billion cubic feet (bcf) of LNG annually.[9]

In Canada, the position was much the same in the mid-2000s. The outlook for LNG import was bullish and as many as nine regasification terminals were proposed at one time (of which Canaport LNG, Rabaska LNG and Cacouna LNG were approved).[10]

We now know that part of the game-changing nature of the growth of shale gas production in North America was to stop a number of the proposed North American regasification terminals in their tracks; as production grew, natural gas prices fell and demand declined (quite the opposite of the previous situation). Of the regasification terminals that are now in commercial operation on the continent, there appears to be significant overcapacity. Figures for 2009 appear to show US and Canadian LNG import capacity standing at around 13 bcf a day;[11] actual imports in 2015 were roughly 0.31 bcf a day.[12]

While it may be fair to say that shale gas production is directly responsible for the overcapacity being experienced by LNG regasification terminals in North America, a number of other factors are also in play.

The tolling model employed by a number of the regasification terminals may well have contributed to the diversion of cargoes originally destined for the United States and Canada to those markets where natural gas prices did not suffer quite the same fate as in North America. The evolution of the value chain from its point-to-point origins has given rise to LNG buyers not necessarily being equity owners of regasification terminals (or having 'in-house' demand for the LNG purchased for delivery to such regasification terminals). Rather, the tolling arrangements appear to have been specifically designed to provide access to what was a lucrative downstream North American trading market for natural gas. As that downstream market has weakened, exacerbated by the fact that there is no global price for natural gas, further recent modifications to the value chain have contributed to diversions, spot trading, flexible shipping arrangements and flexible regasification capacity. While it is impossible to say whether any one of these factors was the key ingredient to the extent of the overcapacity now seen in North American regasification terminals, the new, more flexible supply chain certainly permitted capacity holders or tollers to abandon capacity and seek arbitrage opportunities in different markets.

2.3 The case for LNG export in the United States and Canada

As the North American LNG sector has come to terms with import overcapacity (with only those terminals that alleviate regional bottlenecks generally continuing to import

9 "The Global Liquefied Natural Gas Market: Status and Outlook", Energy Information Administration.

10 "Canadian LNG Import and Export Projects Update", Natural Resources Canada www.nrcan.gc.ca/eneene/sources/natnat/imppro-eng.php. The Canaport LNG, the Repsol and Irving Oil sponsored regasification terminal, was the only terminal to be constructed in Canada, however, and presently supplies regasified LNG principally into the northeastern United States.

11 "Canadian Crude Oil, Natural Gas and Petroleum Products. Review of 2009 and Outlook to 2030", Natural Resources Canada, Petroleum Resources Branch, Energy Sector, May 2011, at p15.

12 Energy Fact Book 2016-2017, Natural Resources Canada, at p55.

meaningful volumes) and shale gas production shows little sign of declining (despite continuing environmental concerns and public interest in this issue), developers have turned their attention to exports. In a short period the value chain has been turned on its head and the United States and Canada are now in the headlines for the number of new LNG export projects proposed and permitted, including, in the case of stranded capacity at the US regasification terminals, the proposal to convert the facilities into 'bi-directional' import-export facilities. Sabine Pass LNG was quickest out of the blocks and has proposed a six-train plant with an export capacity of 27 million tonnes per annum with two trains now in operation and three under construction and expected to be commissioned by 2019. In Canada, Kitimat LNG, which was originally proposed as an import terminal, was the first LNG project to be granted an export licence by the National Energy Board. Since 2011, 24 LNG projects have been issued long-term export licences. The majority of proposed new liquefaction plants have been targeted for Canada's west coast, including Shell Canada's LNG Canada project, Petronas/Progress' Pacific Northwest LNG project and Nexen/Inpex's Aurora LNG project. Six additional projects have been proposed in Quebec and on Canada's east coast.

The abundance of shale gas resources is the key driver for these new LNG export projects; in fact, in the export licence applications filed by Kitimat LNG in Canada and Sabine Pass LNG in the United States (both of which have been granted), the applicants were careful to emphasise to the regulators the overall abundance of natural gas in the market.[13] Sabine Pass LNG even argued that the promotion of an LNG export industry would encourage the development of marginal shale gas resources and increase shale gas production further.[14]

Following the example of Kitimat LNG in Canada and Sabine Pass LNG in the United States, the following projects have made applications for the export of LNG from North America[15] to the Canadian National Energy Board or the US Department of Energy.

Project[16]	**Country**
LNG Canada	Canada (west coast)
Cedar LNG Project	Canada (west coast)
Pacific NorthWest LNG	Canada (west coast)
Prince Rupert LNG	Canada (west coast)

continued on next page

13 Kitimat LNG NEB Licence Application, KM LNG Operating General Partnership, 16 May 2011, p10 and Application of Sabine Pass Liquefaction LLC, supra note 2, p5.
14 Application of Sabine Pass Liquefaction LLC, supra note 2, p5.
15 The North American projects in the table are those which have applied for export licences for LNG to non-US Free Trade Act nations.
16 Information accurate as at end of March 2017. List does not include projects that have been cancelled.

Project	Country
Woodfibre LNG	Canada (west coast)
WCC LNG	Canada (west coast)
Triton LNG	Canada (west coast)
Aurora LNG	Canada (west coast)
Kitsault Energy Project	Canada (west coast)
WestPac Marine Terminal	Canada (west coast)
Steelhead LNG	Canada (west coast)
Grassy Point LNG	Canada (west coast)
Discovery LNG	Canada (west coast)
Orca LNG	Canada (west coast)
Stewart LNG	Canada (west coast)
NewTimes Energy	Canada (west coast)
Malahat LNG	Canada (west coast)
Énergie Saguenay	Canada (Quebec)
Stoltz LNGaz	Canada (Quebec)
Saint John LNG	Canada (east coast)
Goldboro LNG	Canada (east coast)
Bear Head LNG	Canada (east coast)
A C LNG	Canada (east coast)
Freeport LNG Expansion LP	United States

continued on next page

Project	Country
Lake Charles Exports LLC	United States
Carib Energy (USA) LLC	United States
Dominion Cove Point LNG LP	United States
Jordan Cove Energy Project LP	United States
Cameron LNG LLC	United States
Gulf LNG Liquefaction Company LLC	United States
Oregon LNG	United States
Southern LNG Company LLC	United States
Cheniere Marketing LLC	United States
CE FLNG LLC	United States
Commonwealth LNG	United States
Magnolia LNG LLC	United States
Freeport-McMoRan Energy LLC	United States
Venture Global Calcasieu Pass LLC	United States
Advanced Energy Solutions LLC	United States
Argent Marine Management Inc	United States
Eos LNG LLC	United States
Barca LNG LLC	United States
Delfin LNG LLC	United States
Annova LNG Common Infrastructure LLC	United States

continued on next page

Project	Country
Louisiana LNG Energy LLC	United States
Alturas LLC	United States
Strom Inc	United States
SCT&E LNG LLC	United States
Downeast LNG Inc	United States
Air Flow North America Corp	United States
American LNG Marketing LLC	United States
Floridian Natural Gas Storage Company	United States
G2 LNG LLC	United States
Port Arthur LNG LLC	United States
Texas LNG Brownsville LLC	United States
Corpus Christi Liquefaction LLC	United States
Flint Hills Resources LP	United States
Rio Grande LNG LLC	United States
Eagle LNG Partners Jacksonville LLC	United States
Venture Global Plaquemines LNG LLC	United States
Driftwood LNG LLC	United States
Lloyds Energy Group	United States

In aggregate, this amounts to successful and pending applications for licences for export of LNG from North America of more than 102 bcf a day. While it is beyond the scope of this chapter to explore whether the global LNG market could absorb such volumes, what is clear is that even if each proposed project is not built in full, there is an exceptional appetite in North America for the export of LNG.

Perhaps anticipating the level of interest in this sector, and the flood of applications for export licences, in August 2011 the US Department of Energy directed the US Energy Information Administration to analyse the impact of increased natural gas demand for export on the US market, the results of which were delivered in January 2012,[17] and updated upon request in 2014.[18] The Energy Information Administration, unsurprisingly perhaps, concluded in the two studies that increased LNG exports:

- will lead to an increase in domestic natural gas prices;
- will cause natural gas production to increase, with around 75% of such increase coming from shale gas; and
- due to higher prices, will cause part of the extra demand for natural gas required for export to be satisfied by diversion from the power sector and alternative power generation sources to be utilised as a result.

The Energy Information Administration in its Annual Energy Outlook 2017 projects that the United States will be a net exporter of natural gas by 2018, driven by declining pipeline imports and increases in pipeline and LNG exports, and that by 2021 as much as 9.2 bcf of LNG export capacity will be operational.

At present, the Canadian National Energy Board has not engaged in a similar analysis, which might be partly due to historical differences between the US and Canadian energy sectors, with the latter being export-focused long before the advent of an LNG export industry. In addition, the US domestic market for natural gas is bigger than Canada's, so there is greater potential for domestic demand for shale gas in the United States than there is in Canada. However, the greatest challenge facing the Canadian energy sector remains the forecast decline of demand for natural gas from the United States, with LNG export providing the sole means to replace such decline.

Ultimately, and regardless of the regulatory result, the case for LNG export projects in North America will depend on the cost effectiveness of the liquefaction plants and related infrastructure when compared to competing sources of supply around the world. In that respect, the proposed LNG export projects in the United States that are being proposed in the bi-directional model, sharing existing marine and pipeline infrastructure, may well generate efficiencies that enhance such cost effectiveness. Conversely, the proposed Canadian projects' competitive advantage may come from their proximity to Asian destination markets, which may be viewed as the most apt destinations for exports. However, the latter does not appear to be bearing fruit, with no LNG export project in Canada having, at the time of writing, taken a positive final investment decision. There are a number of possible reasons for this and this is discussed further in the next section.

---

17 "Effects of Increased Natural Gas Exports on Domestic Energy Markets", US Energy Information Administration, January 2012.

18 "Effects of Increased Natural Gas Exports on U.S. Energy Markets", US Energy Information Administration, October 2014.

2.4 Destination markets for US and Canadian LNG and project structures

As previously mentioned, the majority of Canadian LNG export projects which have been proposed to date are located on Canada's west coast (in proximity to the Canadian shale gas basins) and are likely to supply the Pacific Basin, due to their obvious proximity and the perceived favourable oil-indexed pricing structures available in those markets.

In the case of the new LNG export projects proposed in the United States, destination markets will likely include South America, Europe, India and (following the Panamax expansion of the Panama Canal) East Asia. Sabine Pass LNG has secured offtake from BG Group, Gas Natural, Kogas, Total and Centrica, demonstrating an even spread between Atlantic Basin and Pacific Basin deliveries. The proposed LNG export project to be developed by Freeport LNG has secured Japanese offtakers for its first train, BP for its second train and SK and Toshiba for train three.

As discussed in Section 2.3, the first wave of new LNG export projects proposed in the early 2010s in the United States are associated with existing regasification terminals, and accordingly will share existing marine, storage and pipeline infrastructure. This will create the first wave of bi-directional LNG facilities in the world and usher in a further evolution in the value chain, moving it further from its original point-to-point model. In this new world, capacity holders will be able to use these new bi-directional terminals to access the US downstream market when natural gas prices are attractive, use the terminal as storage for LNG pending re-export or regasification as prices dictate, or waive regasification capacity altogether when natural gas prices are low, and instead toll low-cost natural gas through the liquefaction trains for export. For such optimisation to be managed effectively, access to flexible transportation arrangements will be key for both pipeline and shipping capacity. This will be the first time that one market will present such significant optionality, being at once a source and destination for LNG.

Meanwhile, the western Canadian experience falls squarely within the traditional value chain which appears to be adopting a point-to-point approach, with the upstream reserve owners owning (through a project vehicle) the liquefaction plant and its capacity, and targeting long-term (Asian) offtakers.[19] This differs from the approach in the United States, where sponsors appear to be adopting a liquefaction tolling model or variations on the same, and this lack of flexibility may be one of the reasons for Canadian projects failing, to date, to have taken a positive final investment decision. Reasons for the divergent approaches are discussed further in Section 3.

2.5 Re-coupling of oil and natural gas pricing?

While the liquefaction and export of shale gas remain in their infancy in North America, the prospect of LNG export prices affecting domestic pricing is unlikely in the short term. If there is significant growth in the volume of LNG exports to markets which tie LNG prices to crude oil, there may be scope for the argument that crude

19 Kitimat LNG NEB licence application, *supra* note 10, pp5-6 and Application by LNG Canada Development Inc, 27 July 2012, pp7-8.

oil prices may once again affect North American natural gas prices. Clearly, there will need to be appropriate flexibility in the value chain to permit diversion of natural gas from domestic consumption to export (and vice versa) to facilitate such price signals to be made. However, given the abundance of shale gas in North America and the pricing methodology proposed in US LNG export projects, this would appear unlikely.

3. Production of shale gas and ramifications for LNG export projects

3.1 Composition

Natural gas from shale formations is similar to natural gas from conventional sources in the sense that its composition or specifications can vary widely from area to area. While natural gas is a mixture of hydrocarbons primarily composed of methane, most natural gas streams will contain smaller amounts of heavier hydrocarbons and other gases or impurities. The percentages of each can vary widely from source to source, and shale gas is no different in this respect. That said, in the face of low natural gas prices the shale gas industry has stepped up production of shale gas resources that produce wet gas to provide a 'liquid boost' to their revenues. In effect, formations which yield a greater volume of natural gas liquids, and which will therefore yield greater crude-oil linked prices for the natural gas liquids on separation, are attractive to producers.[20] This may be a lesser consideration for producers supplying shale gas downstream to an integrated LNG liquefaction plant.

Once extracted, the shale gas is processed to remove other gases, water, sand and impurities, and natural gas liquids may be removed in order for the gas to meet pipeline specifications. The specifications of the inlet gas to a liquefaction plant will need to match the plant's design, and possibly the specification limitations that may be imposed on the LNG seller by the downstream LNG buyer or pipeline specifications at the delivery point. Typically, the liquefaction process will require the prior removal of almost all non-methane components. LNG is usually composed of a very high percentage of methane (well over 90%) as compared to typical natural gas. If LNG is shipped with non-methane components such as natural gas liquids, those components may (depending on the relevant import facility) be removed on receipt.

Typically, the heat content requirements of LNG import terminals in Pacific Basin destination markets are higher than for terminals in Europe and the United States. This means that more natural gas liquids may be left in the LNG than would be the case for LNG destined to other countries.

3.2 Production rates

LNG liquefaction plants rely on a large and stable source of natural gas supply as feedstock and for significant periods of time (typically 20 to 25 years). Shale gas

---

20 It is understood that the price of ethane follows natural gas pricing more closely. However, note the challenges now being faced by the liquids sector and the downward pressure on prices in the face of increased natural gas liquids production and rising inventories.

resources, as dedicated sources of supply, could present some challenges, considering that decline rates for shale gas resources seem to be far more accelerated than those for conventional resources. When looking at a production curve for a shale gas resource as compared to one for a conventional resource, the curve would be significantly steeper and shorter in duration. This could present difficulties for the operator of an LNG liquefaction plant that is reliant on specific shale gas reserves for long-term supply. A related concern is that, partly because the decline rates for shale gas are more difficult to predict than those for conventional reserves, attempts to estimate total recoverable reserves are also much more challenging.

The model outside North America to date is for natural gas to be supplied on a firm and long-term basis, particularly where the purchaser is an LNG liquefaction plant. This has generally presented limited problems, at first blush, for an LNG liquefaction plant's suppliers where the source of natural gas is a conventional field, and clearly this fits well with the value chain downstream where LNG is sold under a long-term, firm, take-or-pay basis.

The challenge for a supplier of shale gas to an LNG liquefaction plant – or the operator of that plant, depending on your perspective – will be offering such long-term firm commitments given the inherent unpredictability of shale gas production volumes over the long term, particularly in the absence of a continued drilling programme designed to bring on more supply from new wells as older wells suffer declines.

Accordingly, the unique production profile of a shale gas resource will affect the contracting strategy that an LNG liquefaction plant will adopt in relation to feedstock supply and the specific market and location of the plant will influence this significantly. For instance, Canada's British Columbia LNG export projects appear to have earmarked northeastern British Columbia's shale gas resources as their dominant source for feedstock; the relative remoteness of the proposed liquefaction plants, together with their potentially constrained pipeline infrastructure,[21] may have an impact on their ability to access alternative sources of supply in the event of production deficits upstream. Conversely, the LNG export projects proposed on the US Gulf Coast have better proximity to a liquid market and pipeline capacity, from which to make good any individual upstream production shortfalls, and the unpredictability of shale gas production rates may therefore be of lesser concern for these projects.

The experience from the Australian coal bed methane to LNG projects will no doubt be of greatest application to the LNG export projects proposed in Canada. As discussed in detail by Gosewisch and Hewitt[22] in the context of the Australian experience, the aggregation of sufficient reserves to ensure long-term supply of shale gas feedstock to an LNG liquefaction plant will present the greatest challenge. It is perhaps for this reason that the sponsors of the larger Canadian LNG export projects proposed to date are, in large part, holders of the upstream reserves that will be exploited to provide feedstock to the liquefaction plants. By taking a more integrated

21 See further discussion on pipeline capacity in Section 3.3.
22 "Coal bed methane for LNG", *The Law and Business of LNG*, (2nd ed) at p94.

and, some might argue, traditional approach to the supply chain in this way, the sponsors of an LNG liquefaction plant will be better able to control production rates and the continued drilling programmes necessary to safeguard those rates. In cases where shale gas supply to an LNG liquefaction plant is on a strictly arm's-length basis (or for those projects where there might be a significant difference in ownership between the upstream production and downstream liquefaction project), the critical risk is that the costs of production and continued drilling programmes for the feedstock supplier in the future do not justify production at the prices available.

The US Gulf Coast LNG export projects appear to be proposed on an entirely different structure, which is reflective of the size and liquidity of US production of natural gas (bolstered of course by increased shale gas production) and the availability of pipeline capacity. Models developed for these projects appear to transfer some or all of the upstream risk onto LNG purchasers, and examples include:

- the use of a liquefaction tolling model, requiring the LNG purchaser to source its own natural gas and, in exchange for a (fixed) capacity and (variable) volumetric fee, to toll such natural gas through the liquefaction plant and receive LNG for offtake; and
- the adoption of a tolling-like approach where feedstock price risk is passed onto the offtaker in full. This arrangement differs from the tolling model described above by keeping the requirement to procure feedstock (and therefore volume risk) with the operator of the LNG liquefaction plant. In this way, upstream risk is passed, in a more limited way, to the LNG offtaker by the direct pass through of the feedstock price as a variable charge in addition to the requirement to pay fixed capacity charges.

In either case, the trade-off would appear to be the elimination or reduction of the take-or-pay commitment on the part of the LNG offtaker in its LNG sale and purchase agreement. As such, LNG offtakers are provided with the flexibility to reduce monthly offtake if the cost of feedstock makes the cost of the LNG uneconomical. These structures may mean that sponsors of an LNG export project are less likely to take an integrated approach to upstream production because they either have no obligation to source supply of feedstock or have limited take-or-pay commitments, and so limited firm demand for feedstock that needs to be secured. This is clearly a model which is unlikely to be replicated in markets that do not share the US Gulf Coast's market liquidity and pipeline capacity characteristics. By requiring offtakers to assume upstream production and/or price risk, these structures require offtakers to venture into newer territory in the LNG value chain, but have the potential advantage of providing those offtakers with LNG that will be priced on a gas-to-gas basis, rather than indexed to crude oil, which appears now to be a key demand from LNG buyers (see, for instance. the basis of the proposed cooperation arrangement between Kogas, Jera and CNOOC). This might be particularly attractive to LNG offtakers, with Henry Hub prices at their current low levels. In addition, Japanese LNG buyers which have accounted for a significant portion of global LNG demand are focused on securing flexibility from LNG sellers due to pressures now coming from proposed or newly liberalised domestic energy markets in Japan.

The liquefaction tolling model also assumes that incremental shale gas production will continue to come online over the medium to long term to maintain liquidity in the market; in the case of shale gas, it is reasonable to assume that this would require natural gas prices to be at a level that incentivises such continued investment in the sector.

3.3 Location of shale gas plays relative to LNG facilities

Given the necessary coastal location of LNG liquefaction, storage and loading facilities, shale gas reserves that are closer to such coastal locations are obviously preferable from the point of view of the sponsors of an LNG export project. The locations of the shale gas reserves and the facilities will dictate the necessary infrastructure required to bring the shale gas to the liquefaction plant. With respect to the Montney and Horn River shale gas areas of northeast British Columbia (which would be the sources of gas for most of the proposed Canadian west coast LNG export projects), the Canadian National Energy Board has stated that "there is not enough infrastructure in northeast [British Columbia] to handle growth in shale gas production beyond the next few years".[23] There have since been a number of pipeline proposals:

- the 463 kilometre (km) Pacific Trail Pipelines project, connecting the Enbridge (formerly Spectra) British Columbia Pipeline System to the proposed Kitimat LNG project in Kitimat;
- the 700km Coastal Gas Link project, connecting the Montney producing area to the proposed LNG Canada project in Kitimat;
- the 850km Spectra Energy and BG Natural Westcoast Connector Gas Transmission project, connecting Fort St John to the proposed BG Group LNG project in Prince Rupert (although Shell has indicated that this LNG export project will not be proceeding);
- the twinning of the existing 525km Pacific Northern Gas pipeline between Summit Lake, British Columbia and Kitimat; and
- the 900km (780km terrestrial and 120km marine) Prince Rupert Gas Transmission project to be operated by TransCanada Corporation on behalf of Progress Energy, connecting Progress' operations in the Montney to its Pacific NorthWest LNG project.

Presently, the pipelines proposed above are each anticipated to supply a single LNG export project, with the exception of the Pacific Northern Gas pipeline. The terrain in northeast British Columbia is rugged, and while the area does contain large areas of crown land which the British Columbia government may well lease for pipeline siting, there are other legal issues involved with the development of what are, in their own rights, significant infrastructure projects, without considering the LNG liquefaction plants to which they will be connected. Jurisdiction for the permitting of these pipelines lies in most cases with the British Columbia Oil and Gas Commission (not the National Energy Board which grants export licences, unless the pipeline also

---

23 "A Primer for Understanding Canadian Shale Gas", National Energy Board of Canada Energy Briefing Note, November 2009.

crosses an inter-provincial or national border). As part of this permitting process, public and Aboriginal consultation will be required. The conduct of the permitting process will be viewed against the backdrop of significant interest in the region with respect to new pipelines. The Northern Gateway Pipeline (a crude oil pipeline) was approved by the National Energy Board and Federal Government before having such approvals quashed by the Supreme Court due to deficiencies in the Federal Government's Aboriginal consultation process, and the Trans Mountain Pipeline (another crude oil pipeline) has also been approved but is now subject to numerous legal challenges. Given the number of US projects in operation or construction as compared to Canada, this appears to be another key driver for timing. None of the proposed natural gas pipelines listed above is currently being constructed.

The LNG export projects proposed in the United States do not presently require such extensive pipeline development, given in part that in many cases, those projects will be located side by side with existing regasification terminals and share existing pipeline infrastructure in large part. This may well present significant cost savings for these projects, as well as reducing overall project development timelines.

4. Financing shale gas LNG projects

4.1 Financing in Canada

Financing for proposed LNG export projects may take different forms dictated by the nature of the project sponsor and the proposed project structure, including integrated project financing of various facilities in the value chain or separate financing of individual facilities. Some participants may use corporate equity or debt financing for upstream facilities and access commercial credit for downstream facilities, while others may use commercial credit for all of the facilities.[24]

In the past, access to Canadian project finance debt has been somewhat limited as Canadian banks have been reluctant to provide long-term debt. Thus, Canadian infrastructure projects often received long-term financing led by foreign lenders with some Canadian bank participation, or more often long-term financing from a limited number of Canadian life insurance companies. More recently, the pool of Canadian providers of long-term debt has increased significantly and Canadian infrastructure projects have been financed by a combination of short-term bank debt and long-term project bonds. While Canadian banks have become more competitive for short-term debt, long-term debt availability remains a challenge without strong bank relationships.

As the Canadian project bond market becomes more competitive, more infrastructure projects are being financed by project bonds. Canadian and foreign institutional investors have provided long-term bond financing for several Canadian infrastructure projects. Financing for any proposed LNG export projects in Canada will likely come from a combination of foreign banks, as well as foreign and Canadian institutional investors, with potential support from Export Development Canada.

In the United States, Sabine Pass LNG has closed credit facilities that will be used to fund the first five trains of its LNG export project. They appear to be classic project

24 See also Douglass at p203, *supra* note 18.

financings. The project structure for Sabine Pass LNG appears to be based, at least in part, on the liquefaction tolling model previously discussed. Project lenders have clearly indicated a comfort level with that structure, given the successful financial close of the financing, but it is unlikely that the proposed Canadian LNG export projects will be able to adopt a similar structure for the reasons discussed in Sections 2 and 3.

4.2 Project risk

Foreign lenders in North America will enjoy the same property rights as citizens. The risk of nationalisation or government expropriation of assets is considered extremely limited. Any political risk related to an LNG export project is more likely to be related to individuals or groups that exercise property or other rights or seek to shape public opinion in an attempt to halt a project. As LNG export projects require different levels of government approval, elected government officials may decide to abandon support for a project in the face of great public opposition.

In addition to the foregoing risks, LNG export projects requiring the supply of feedstock from shale gas will involve additional project risks. For example, the lenders will require independent confirmation of the shale gas reserves to ensure that projected feedstock volume is sufficient to satisfy the commitments made to the LNG offtakers – as discussed in Section 3.2, difficulties in determining decline rates and recoverable reserves will be key project risks for lenders. Given that the level of reserves may change or may need to be proved as the project proceeds, the potential reduction of the estimated reserves for a shale gas LNG export project presents additional risk for lenders. In mitigation of such risk is the fact that unlike the vast majority of liquefaction plants around the world, plants in the United States and Canada will have access to an interconnected pipeline network and should be better able to manage production declines or periodic shortfalls in the supply of feedstock. In any event, it would appear that this project risk is one that lenders to the Canadian LNG export projects will have to grapple with the most, given the trend in the United States towards the liquefaction tolling model.

Lenders must also address risks related to the projected LNG pricing. As discussed above, shale gas production has caused domestic natural gas pricing to fall, making it important that the LNG pricing for export purposes remain high enough for the project to service its debt. LNG price formulae pegged to domestic North American hub pricing may give lenders some concerns. In the case of Kitimat LNG, destination market-based price formulae appear to be the approach that will be proposed to LNG offtakers, but whether that is palatable for offtakers remains to be seen – so far that would not appear to be the case. Once again, under the liquefaction tolling model, price risk is assumed by the LNG offtaker/toller and may be of lesser concern to the lenders. What will be of greater concern is the size of the fixed capacity payments payable by the offtaker/toller and its creditworthiness.[25]

25 Kitimat LNG NEB Licence Application, *supra* note 10, p7.

About the authors

Yanal Abul Failat
Associate, LXL LLP
yaf@lawxl.com

Yanal is an international lawyer (qualified as a solicitor in England and Wales) specialising in energy law at LXL LLP, advising state-owned companies, utilities and other major energy corporations on general corporate and commercial operations, M&A, regulatory matters and dispute resolution.

He principally acts on matters in the oil and gas sector, and has experience in oil and gas regulation and contracting across the entire supply chain (upstream, midstream and downstream). Yanal has advised exploration and production companies operating in the UK Continental Shelf and elsewhere in Africa, Asia, Europe and the Middle East.

Scot Anderson
Partner, Hogan Lovells
scot.anderson@hoganlovells.com

Scot Anderson is the global head of the energy and natural resources group for Hogan Lovells, and counsels and supports clients on nearly every legal issue affecting the extractive industries, both domestically and internationally. He oversees complex transactions for natural resources clients and represents clients in administrative proceedings, as well as in federal and state courts. He regularly addresses issues related to mineral development of federal public lands and Indian lands. Scot also advises clients on international project development and transactions. Scot began his legal career with the Denver office of an Am Law 100 firm and then served 10 years as in-house counsel for Atlantic Richfield Company (ARCO) in Denver and England. Scot received his PhD from the University of Colorado in 1989. He earned his JD in 1987 and his MA in 1982 from the University of Iowa. He received his BA from Augustana College in 1980.

Vivek Bakshi
Partner, Dentons Canada LLP
vivek.bakshi@dentons.com

Vivek Bakshi is a partner in the energy group at Dentons Canada LLP, one of Canada's leading natural law firms, and is based in its Toronto office. Mr Bakshi focuses on M&A and project development in the energy, natural resources and infrastructure sector. He has particular expertise in the liquefied natural gas (LNG) sector working with LNG traders and sponsors in connection with the trade of LNG and development of LNG liquefaction plants and regasification terminals worldwide. Mr Bakshi has worked in London, Singapore and Tokyo prior to practice in Canada, and speaks regularly around the world on matters of energy law and practice. Mr Bakshi is qualified to practise Ontario law and is also a member of the Law Society of England and Wales.

Robert A Bassett
Mining team leader, Holland & Hart LLP
rbassett@hollandhart.com

Robert A Bassett is the mining team leader at Holland & Hart LLP in Denver, Colorado, a full-service law firm with 16 offices in the western United States and in Washington, DC, and one of the largest practice groups focused on mining law in the nation. He practices natural resources and international law with an emphasis on the mining industry. In recent years he has handled projects as diverse as the financing of coal, molybdenum, uranium and copper mines and companies, in litigation involving coal mines and pipelines, and in mergers and acquisitions of Canadian and US mining companies.

Mr Bassett has published numerous articles for the Rocky Mountain Mineral Law Foundation, of which he is the International Committee chairman, and for the International Bar Association Section on Energy and Natural Resources Law, of which he is coordinator of the Model Mine Development Agreement Project and former chair of the Mining Committee.

Michael Darowski
Partner, Gowling WLG
michael.darowski@gowlingwlg.com

Michael Darowski is a partner in Gowling WLG's disputes practice in London. Michael's practice focuses on energy and natural resources disputes, handling a broad range of complex, high-value international commercial arbitrations and litigation. Michael acts for national and international oil companies in resolving disputes involving English and many other governing laws, and has acted in arbitrations under the rules of all of the main arbitral institutions and in *ad hoc* disputes. He has been involved in national court proceedings in various jurisdictions in relation to all aspects of the arbitration process. Michael has written articles and spoken about energy disputes at industry conferences, including on the development of shale gas in Europe and the disputes that are likely to arise.

Jessica Davies
Counsel, Clifford Chance
jessica.davies@cliffordchance.com

Jessica Davies is an international oil and gas lawyer in Clifford Chance's Perth office. Jessica has extensive experience advising multinational organisations on all aspects of the oil and gas supply chain, including project development, joint operating agreements, LNG sale and purchase agreements, gas transportation, processing and marketing agreements, M&A, and operational and strategic issues.

Jessica has practised in Australia, New York and London. She has previously received AMPLA's inaugural Australian energy and resources lawyer award, based on her demonstrated commitment, achievement and international expertise, and spoke at the Rocky Mountain Mineral Law Foundation's International Law Conference in the United States.

Steve Delic
Counsel, BP Exploration and Production – Australia
steve.delic@bp.com

Steve Delic is an oil and gas lawyer for BP in Perth. He advises BP on all aspects of its upstream projects in Australia, from exploration through to developments, operations and marketing. Prior to joining BP, Steve was a senior associate with a major Australian law firm and then worked as in-house counsel for Shell.

Steve has expertise on large-scale integrated gas projects and gas value chains. In particular, he has extensive experience advising on procurement, joint ventures, M&A and LNG/domestic gas matters, including disputes, regulatory issues, tolling transactions, equity lifting arrangements, and sale and purchase agreements.

Joseph Dutton
Associate research fellow, Energy Policy Group, University of Exeter
j.dutton@exeter.ac.uk

Joseph Dutton is an associate research fellow at the Energy Policy Group (University of Exeter). His research focuses on the politics of European energy policy, electricity market integration and interconnection, and the evolution of European energy governance. He has also researched the globalisation of UK gas supply, the development of shale gas in the United Kingdom and Europe, and the changing dynamics of global gas markets. Previously he worked as a commodities price reporter covering the UK and northwest European natural gas markets, and as an analyst in upstream oil and gas consultancy. Joseph holds an MA in international relations and European studies from the University of Kent.

Sharon Eacott (Wilson)
Alinta Energy
Sharon.eacott@alintaenergy.com.au

Sharon is the general counsel and company secretary of Alinta Energy. Prior to joining Alinta, she was a partner with Herbert Smith Freehills specialising in the energy sector. Sharon has over 15 years' experience advising on Australia's largest energy project, as well as energy and resources M&A, project development and operations across the Middle East, South East Asia and Latin America.

Belinda Fan
Partner, Herbert Smith Freehills
belinda.fan@hsf.com

Belinda Fan is a partner in the corporate group of Herbert Smith Freehills, based in Melbourne.

Belinda's practice focuses on the infrastructure, energy and resources industries. She advises on project structuring, acquisitions and divestments, upstream and midstream developments, and downstream operational and commodity supply arrangements. She also specialises in advising foreign investors on Australia's foreign investment regulations.

Belinda's experiences include acting on gas supply arrangements and M&A transactions relating to Queensland's LNG projects. She also recently advised on the development of the Northern Gas Pipeline connecting the Northern Territory to the East Coast of Australia.

Belinda is a fluent Mandarin speaker. She holds degrees in law and commerce, and is a member of the Association of International Petroleum Negotiators and the Australian Institute of Company Directors.

Nicholas Fulford
Global head of gas and LNG, Gaffney, Cline & Associates
nick.fulford@gaffney-cline.com

Nicholas Fulford leads the gas and LNG practice at Gaffney, Cline & Associates (GCA). He coordinates global gas activity across the company's main offices situated in London, Houston and Singapore. He is a chartered engineer, with a Master's degree from the Royal Military College of Science.

Nick has over 35 years' experience in the gas value chain, having held senior roles in the past at British Gas/Centrica covering gas production, gas contract negotiation and LNG shipping, and having established a North American business for Centrica, with a focus on gas and power wholesale and retail markets, and unconventional gas resource development.

Since joining GCA in 2013, Nick has assisted a range of clients in gas monetisation, pricing and risk management, and gas master planning, as well as due diligence and technical assessment of gas projects globally.

Andrew Lillie
Partner, Hogan Lovells
andrew.lillie@hoganlovells.com

Andrew Lillie heads the Litigation Practice of Hogan Lovells in the Denver office. He is a commercial litigator and regulatory attorney who focuses on energy and natural resources businesses and related commercial and environmental litigation, regulation, and compliance. He has also served as a Special Assistant Attorney General for Colorado agencies, and was a law clerk for Chief Judge Lewis T Babcock of the United States District Court for the District of Colorado.

In 2002, Andrew received both his JD, *Order of St. Ives*, from the University of Denver Sturm College of Law and his MA from the University of Colorado. He earned his BGS from the University of Michigan College of Literature, Science, and the Arts and his BS, with distinction, from the University of Michigan School of Natural Resources and Environment in 1994.

Jennifer Morrissey
Counsel, Dentons US LLP
jennifer.morrissey@dentons.com

Jennifer Morrissey is counsel in the global energy practice group in the Washington, DC office of Dentons US LLP. She holds a JD from Georgetown University and a Master's degree from Johns Hopkins University. She advises a variety of clients in the natural gas and electricity sectors, including project developers, producers, end users, regulators and financial institutions. Her practice includes administrative and appellate litigation, as well as regulatory and transactional matters. She has broad litigation experience in traditional gas and electric rate proceedings before the Federal Energy Regulatory Commission, enforcement actions, and water rights, among other matters.

Her transactional experience ranges from negotiating simple agreements to advising on substantial investments in energy sector assets. She frequently speaks before industry associations and graduate students, and has authored dozens of articles in the energy press on energy policy and regulation, many of which focus on the natural gas industry.

Ryan Pereira
Principal commercial manager – global gas & LNG, Gaffney Cline & Associates
ryan.pereira@gaffney-cline.com

Ryan Pereira is principal commercial manager in the global gas & LNG group at Gaffney Cline & Associates (GCA). Based in London, Ryan has extensive international experience and cultural awareness having completed assignments across Europe, the Middle East, Africa, Asia Pacific, North America and the Caribbean across the downstream, chemicals, midstream and LNG, and upstream sectors. He holds a degree in economics and qualifications in management accounting, corporate finance and treasury, and project management.

Ryan has over 14 years of experience in the energy sector, from industry and advisory roles with majors (ExxonMobil) and operators (Centrica Energy), as well as working with joint venture parties and national oil and gas companies.

Since joining GCA in 2014, Ryan has led projects covering LNG import and exports, gas to power, M&A, gas disputes and arbitrations, gas market studies, energy market reform, and delivery of LNG training, and has written several articles for external publications covering global gas markets.

Elizabeth Titus
Counsel, Hogan Lovells
liz.titus@hoganlovells.com

Elizabeth (Liz) Titus represents clients in the energy and natural resource industries. Liz's practice focuses on the extractive industries, particularly oil and gas (upstream and midstream), and mining. Liz supports clients with projects and operations on both private and public lands. She negotiates commercial agreements with purchasers, landowners, royalty holders and other business partners. Liz is also a trained litigator and has trial court, appellate and administrative hearing experience. Her disputes work has covered the processing, sale, and transportation of oil, natural gas and related products; rights to surface use; and payment of royalties. Liz's clients are subject to extensive federal, state and local regulation; she helps them comply at all levels of government. Liz earned her JD, Order of St Ives, from the University of Denver Sturm College of Law in 2006. She received her BA, *cum laude*, from Smith College in 1999.

Index

References to Figures are in *italic* print; 'LNG' is short for 'liquefied natural gas', whereas 'SPAs' stand for 'sales and purchase agreements'.